SpringerBriefs in Energy
Energy Analysis

Series Editor: Charles A.S. Hall

For further volumes:
http://www.springer.com/series/10041

Jeremy J. Wakeford

Preparing for Peak Oil in South Africa

An Integrated Case Study

 Springer

Jeremy J. Wakeford, Ph.D.
Senior Lecturer Extra-ordinary
School of Public Leadership
Stellenbosch University
and
Chairman
Association for the Study of Peak Oil South Africa
jeremy@aspo.org.za

ISSN 2191-5520   ISSN 2191-5539 (electronic)
ISBN 978-1-4614-9517-8   ISBN 978-1-4614-9518-5 (eBook)
DOI 10.1007/978-1-4614-9518-5
Springer New York Heidelberg Dordrecht London

Library of Congress Control Number: 2013953548

Printed on acid-free paper

Springer is part of Springer Science+Business Media (www.springer.com)

Preface

I first encountered the concept of "peak oil" in 2003 and was immediately struck by its profound implications for my academic discipline—economics—as well as for society more generally and especially for South Africa where I live. It was only 2 years later that I decided to pursue peak oil as my main area of research. At the time, very few people in South Africa seemed to know about the Hubbert peak, and the notion of resource limits was a taboo in mainstream economics, not to say society more generally. Fortunately, I soon became part of a small but dedicated network of people who shared my interest and concerns about peak oil. This book is the culmination of 8 years of research on peak oil and its implications for South Africa. It is a condensed version of my doctoral dissertation, which was accepted at Stellenbosch University in May 2012. The main message of the book is this: if we continue along a business-as-usual pathway of oil dependence, then we are destined for socioeconomic shocks and hardship; but if we proactively implement measures to mitigate the varied impacts of peak oil then we will be using this opportunity as a catalyst for creating a more sustainable future for our society and the natural environment we depend upon.

This book is written primarily for policy makers, government officials, academics, planning authorities, and business leaders in the fields of energy, transport, agriculture, economics, and social development. These are the people who have an opportunity to influence the development path of their country. But I hope that it will also appeal to and inform citizens concerned about our collective future, as political leaders often need prodding from their constituents. Although the book is largely a case study of South Africa, I firmly believe that it is of relevance to many other nations and to the world at large. This is because South Africa represents in many ways a microcosm of our global society—an uneasy mix of sophisticated industries and survivalist economies, wealthy elites and poverty-stricken millions. If this nation could wake up to the realities of an oil-constrained future and implement a range of mitigation strategies, it could serve as a model for the world on how to undertake a difficult socioeconomic transition from the fossil fuel era to a more sustainable regime. A key question is how our remaining stocks of fossil fuels can be utilized most effectively to enable this transition.

The bulk of the book is organised thematically according to five main subsystems of the South African socioeconomic system, namely, energy, transport, agriculture, economy, and society. Chapters 2–6 include a brief overview of each subsystem, an analysis of its oil dependencies, a discussion of the likely impacts of peak oil under a business-as-usual policy context, and a presentation of recommended mitigation strategies, policies, and measures. Chapter 7 places these policy recommendations within a broader conceptual framework informed by the literature on societal transitions to sustainability. This concluding chapter presents a preliminary vision of a post-oil, sustainable future, along with a transition action plan as a guide to its attainment, but also discusses the obstacles and risks facing implementation of such a transition plan. Finally, the chapter highlights some important areas for deepening the research.

I am grateful to many people for contributing in various ways to the underlying research and to the ultimate completion of this book. I owe an intellectual debt to the many pioneering authors—too many to mention individually—whose work I have drawn upon and whose ideas have helped to shape my own understanding of peak oil and its implications. The Centre for Renewable and Sustainable Energy Studies at Stellenbosch University provided a bursary which enabled my doctoral research. My PhD supervisor, Professor Mark Swilling, bravely took on what seemed at times like an impossibly broad topic and generously shared his insights, knowledge, and expertise in the area of sustainability. Fellow members of Professor Swilling's doctoral research group gave me feedback and encouragement at several research colloquiums, and my PhD examiners provided helpful suggestions for improvement of the dissertation. My fellow members of the Association for the Study of Peak Oil South Africa (ASPO-SA) have since 2006 provided an invaluable intellectual peer group, generating many stimulating discussions about peak oil and its implications for our country and the world as well as constructive criticism of many of my shorter articles.

On a more personal note, I am grateful to my mother and my late father for providing me with a foundation of unconditional love, support, and encouragement in all my endeavours and for providing me with the best possible educational opportunities at school and university. My daughter Jade is a constant source of inspiration and delight, with her inquiring mind, creative imagination, and appreciation of life. Above all, I thank my wife Jacqui for providing moral support, for sharing my passion for peak oil research, for being ever-willing to debate ideas, and for diligently proofreading my doctoral dissertation and the book manuscript.

Finally, I wish to acknowledge the friendly and efficient editorial support of the Springer team, led by David Packer and Sara Kate Heukerott, and the high-quality input and encouragement of series editor Professor Charles Hall.

Cape Town, South Africa   Jeremy J. Wakeford

Contents

List of Abbreviations and Acronyms

ANC	African National Congress
APPGOPO	All Party Parliamentary Group on Peak Oil
ASPO	Association for the Study of Peak Oil
BEV	Battery electric vehicle
BRT	Bus rapid transit
C&D	Cap and Dividend
C&S	Cap and Share
CAV	Compressed air vehicle
CHP	Combined heat and power
CNG	Compressed natural gas
CPI	Consumer price index
CRDP	Comprehensive Rural Development Programme
CSIR	Council for Scientific and Industrial Research
CSP	Concentrated solar power
CTL	Coal-to-liquid
CWP	Community Work Programme
DAFF	Department of Agriculture, Forestry and Fishing
DCoGTA	Department of Cooperative Governance and Traditional Affairs
DEA	Department of Environmental Affairs
DEAT	Department of Environmental Affairs and Tourism
DME	Department of Minerals and Energy
DoE	Department of Energy
DoT	Department of Transport
DPLG	Department of Provincial and Local Government
DPW	Department of Public Works
DRDLF	Department of Rural Development and Land Reform
DTI	Department of Trade and Industry
EDD	Economic Development Department
EIA	Energy Information Administration
EPWP	Expanded Public Works Programme
EROI	Energy return on investment

FAO	Food and Agriculture Organisation
FSSA	Fertilizer Society of South Africa
GCIS	Government Communication and Information Systems
GCV	Grid-connected vehicle
GDP	Gross domestic product
GHG	Greenhouse gas
GTL	Gas-to-liquid
GVA	Gross value added
HEV	Hybrid electric vehicle
ICE	Internal combustion engine
ICEV	Internal combustion engine vehicle
ICT	Information and communication technology
IDC	Industrial Development Corporation
IDP	Integrated Development Plan
IEA	International Energy Agency
IMF	International Monetary Fund
IPAP	Industrial Policy Action Plan
IRP	Integrated Resource Plan
LED	Local economic development
LNG	Liquefied natural gas
LPG	Liquid petroleum gas
MBT	Minibus taxi
NATMAP	National Transport Master Plan
NDA	National Department of Agriculture
NFLED	National Framework for Local Economic Development
NFSD	National Framework for Sustainable Development
NGL	Natural gas liquids
NGP	New Growth Path
NMT	Non-motorised transport
NPC	National Planning Commission
NSDP	National Spatial Development Perspective
NSSD	National Strategy for Sustainable Development
NTAP	National Transition Action Plan
NHTS	National Household Travel Survey
OCGT	Open-cycle gas turbine
OECD	Organisation for Economic Cooperation and Development
OPEC	Organisation of Petroleum Exporting Countries
ORTIA	O.R. Tambo International Airport
PAP	Pollution authorisation permit
PHV	Plug-in hybrid vehicle
PRASA	Passenger Rail Authority of South Africa
PV	Photovoltaic
RE	Renewable energy
REM	Rare earth metal
RSA	Republic of South Africa
RTMC	Road Traffic Management Corporation

SA	South Africa
SAPIA	South African Petroleum Industry Association
SARB	South African Reserve Bank
SARCC	South African Rail Commuter Corporation
SAR&H	South African Railways & Harbours
SAWEA	South African Wind Energy Association
StatsSA	Statistics South Africa
TDM	Travel demand management
TEQ	Tradable energy quota
TIPS	Trade and Industrial Policy Strategies
TFEC	Total final energy consumption
TPES	Total primary energy supply
UA	Urban agriculture
UCG	Underground coal gasification
UK	United Kingdom
US	United States
VAT	Value-added tax
WEO	World Energy Outlook
WTO	World Trade Organisation

Units of Measurement

bbl	Barrels
boe	Barrels of oil equivalent
bpd	Barrels per day
Btu	British thermal unit
Gb	Gigabarrels (billion barrels)
GJ	Gigajoule
Gt	Gigatonnes (billion tonnes)
GW	Gigawatts (billion watts)
GWh	Gigawatt hours
kbpd	Thousand barrels per day
km	Kilometre
km/h	Kilometres per hour
kt	Kilotonnes
kWh	Kilowatt hours
mbpd	Million barrels per day
mBtu	Million British thermal units
MJ	Megajoules (million joules)
mlpa	Million litres per annum
Mt	Megatonnes (million tonnes)
MW	Megawatts (million watts)
MWh	Megawatt hours
R	Rands
Tcf	Trillion cubic feet

Energy Conversion Factors:

1 Joule $= 9.48 \times 10^4$ Btu.
1 mBtu $= 1{,}055$ MJ.
1 GWh $= 3{,}600$ GJ.
1 boe $= 5.45 \times 10^6$ Btu.

List of Figures

List of Tables

Chapter 1
Introduction: The End of Cheap Oil and Its Implications for South Africa

Oil is the master resource that fuels the world economy, providing 33 % of global primary energy supply, meeting over 40 % of final energy demand, supplying 95 % of the energy powering global transport systems, and providing feedstock for the diverse petrochemicals industry (IEA 2013). For over a century, and especially since the Second World War, growth in the world economy has been strongly correlated with growth in oil consumption (Hall and Klitgaard 2012; Hirsch 2008). Oil has impressively boosted agricultural productivity and thus allowed a massive expansion of the world's human population from two billion in 1930 to over seven billion today (Brown 2008). Cheap transport fuels derived from oil have enabled the globalisation of the world economy and the urbanisation of half of humanity (Rubin 2009). The International Energy Agency (IEA 2012) has forecast that global demand for oil could grow by 14 % by 2035, with all of the net additional demand projected to come from emerging economies. This rise in demand is expected to be driven almost entirely by increasing use of motorised transport for both passengers and freight as incomes rise in developing countries. This forecast is however premised on the assumption that the world economy will grow by an average annual rate of 3.5 % over the period and that oil prices will reach only $125 (in 2011 dollars) by 2035. However, there are many reasons to doubt these assumptions (Wakeford 2012).

1.1 Peak Oil in a Nutshell

Since the late 1990s, a growing body of literature by academics and oil industry experts has been warning that the historical trend of increasing supplies of oil cannot continue indefinitely (Sorrell et al. 2010a, b). This is because oil (like other fossil fuels), having been formed in the geological past, is a finite resource subject to depletion (Aleklett and Campbell 2003). This finiteness necessarily implies that at some point in time, the annual rate of production of oil at a global scale must reach an all-time maximum and begin an irreversible decline (Hubbert 1956). This "peak oil" phenomenon, as it is commonly termed, has already been observed to occur in the majority of individual oil-producing countries and in large regions such as North America and Europe (Hirsch 2008; Sorrell et al. 2010b). Evidence suggests that the world is nearing the global oil production peak. Global new oil discoveries reached a maximum in the 1960s and have been on a declining trend ever since, despite remarkable improvements in exploration, drilling, and extraction technologies and record high prices in recent years (ASPO Ireland 2009; Benes et al. 2012). Data from the US Energy Information Administration (EIA 2013a) show that conventional crude oil production—oil from wells accessed using typical drilling techniques—has been essentially flat at around 74 million barrels per day (mbpd) since 2005. The International Energy Agency has recently stated that conventional crude peaked in 2008 (IEA 2012). Whether or not the world has precisely reached a peak, it seems clear that production can no longer increase at historical rates of over 3 % a year.

Much of the new oil that has come on stream in recent years has come from unconventional sources such as Venezuela's heavy oil, Canada's oil sands, and America's shale or "tight" oil, all of which require special—and costly—extraction and refining techniques. Although the resource estimates for unconventional oil are generally large, the annual flow rate of production is constrained by a number of economic, physical, and environmental factors (Aleklett et al. 2010). Firstly, the hugely capital-intensive nature of these production processes means that marginal production costs for unconventional oil are much higher than those of conventional oil and that there are physical limits to the number of wells that can be drilled in the medium term. Secondly, the environmental impacts of unconventional oil production are significantly greater than those of conventional oil: the freshwater demands are much greater, the CO_2 emissions can be up to twice as high per barrel of oil, and hydraulic fracturing and oil sand production may pollute freshwater sources (Hughes 2011). Two recent reports based on assessments of historical data for thousands of producing wells in the United States suggest that the much-vaunted shale or "tight" oil boom might turn out to be a 10-year bubble (Hughes 2013; Zittel et al. 2013).

Thus while unconventional oil production has allowed total world oil production to continue to expand slowly in recent years while conventional oil output has stagnated, they have come with a substantially higher economic and environmental cost. Furthermore, according to some independent researchers these unconventional sources are unlikely to offset the depletion of conventional oil production for more than a few years (Aleklett et al. 2010). Many analysts are warning that the global production of all types of oil (both conventional and unconventional) might begin to decline this decade (Hirsch et al. 2010; Sorrell et al. 2010a). The post-peak rate of decline in oil production could be between 2 and 5 % per annum, depending on a complex combination of geological, economic, and political factors (Hirsch 2008).

For net oil-importing nations such as South Africa, and for international crude oil prices, the quantity of oil traded on international markets is of even more immediate significance than total world oil production. Data from the US Energy Information Administration (EIA 2013a) show that world oil exports have stagnated since 2005, which largely explains the steep rise in oil prices in recent years (see Fig. 1.1). It is highly likely that world oil exports have passed their all-time peak because domestic consumption of oil is on a rising trend in most oil-exporting countries, driven by growing populations and/or rising incomes (Brown 2013).

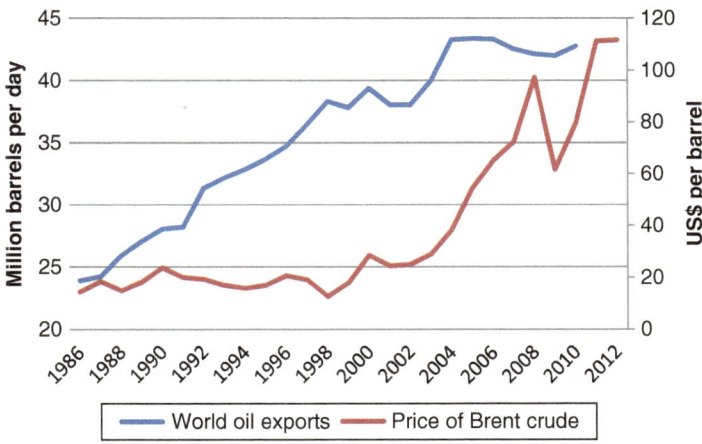

Fig. 1.1 World oil exports and crude oil price, 1986–2012. *Source*: BP (2013); EIA (2013a)

Not only is the quantity of oil available on world markets becoming scarcer, but the quality of available oil is also deteriorating. This is principally because the easier to access deposits of "light sweet" crudes, typically discovered decades ago, are being steadily depleted and the frontier for new oil has moved into more remote areas such as deep offshore wells, polar regions, and unconventional oil sources, which are economically more costly and technically more difficult to access and process (Gagnon et al. 2009). Thus the energy return on (energy) investment (EROI) for oil, which measures the ratio of energy delivered by the process of oil exploration and extraction to the energy inputs, is diminishing in the world as a whole and in most individual countries (Guilford et al. 2011). Furthermore, the EROI for unconventional oil resources such as oil sands and shale oil is estimated to be less than 5:1 (Murphy and Hall 2010). Thus the net energy surplus (i.e. the energy output minus the energy input) yielded by oil is set to decline at a faster rate than the gross quantity of oil produced; and this will put further upward pressure on oil prices.

Conventional economists assume that there will be adequate substitutes for oil that will become economically competitive when oil prices are sufficiently high. An alternative perspective founded on biophysical principles, in contrast, draws attention to the vital quality characteristics of oil and alternative energy sources, especially the EROI (Hall and Klitgaard 2012). Oil has several special characteristics that explain its position as the world's foremost energy source, including historically high EROI, massive energy surplus, ease of storage and transport, and versatility of use (Cleveland 2008). Although there are many potential substitutes for oil, all of them have important quality limitations, such as intermittency, limited scale and/or concentration, low EROI, and in some cases harmful environmental impacts (Heinberg 2009). Thus alternatives may not be more economically *viable* even with higher oil prices. Furthermore, it will take considerable amounts of time and investment capital to develop and scale up the alternatives, while current infrastructures and spatial development ensure "lock-in" to oil dependency for many years (Hirsch et al. 2005). There is a strong possibility that the total net energy accessible globally will decline once oil production begins its descent, and this could be compounded within a decade or two by declines in coal (Mohr and Evans 2010; Patzek and Croft 2010) and conventional nuclear energy supplies (Dittmar 2013; Zittel et al. 2013). Enhanced energy efficiency will be necessary but is subject to diminishing returns and the "rebound effect", whereby income savings generated from efficiency can lead to increased expenditures on energy both directly and embedded in goods and services (Berkhout et al. 2000).

There is a chance, of course, that one or more technological breakthroughs in the fields of energy and transport may occur, which in time could revolutionise mobility and reduce demand for oil. At present, the most popular contender is unconventional natural gas made accessible by hydraulic fracturing. In a recent report, the US Energy Information Administration estimated that world shale gas resources could amount to 7,200 trillion cubic feet (tcf) or 32 % of total unproved gas resources (EIA 2013b). However, this figure is highly speculative and is based on uniquely favourable circumstances pertaining in the United States. The economically viable quantity of gas that might be produced from shale basins in other parts of the world could be significantly smaller due to less favourable geological conditions, shortages of drilling equipment and skilled personnel, and environmental concerns (Zittel et al. 2013).

Another wildcard is low-energy nuclear reaction (LENR) processes, formerly known popularly as "cold fusion", which apparently involve excess energy produced from reactions of hydrogen with nickel or palladium (Kim 2013; Srivastava et al. 2010). According to the chief scientist at the US National Aeronautics and Space Administration (NASA) Langley Research Center, Denis Bushnell, "we now have over two decades of hundreds of experiments worldwide indicating heat and transmutations with minimal radiation and low energy input" (Bushnell 2012). Bushnell contends that "there are estimates using just the performance of some of the devices under study that 1 % of the nickel mined on the planet each year could produce the world's energy requirements at the order of 25 % the cost of coal". However, it is too early to tell whether LENR energy can be successfully commercialised on a large scale, and the investments required would be enormous and difficult in a capital-constrained world. Even if such a technology were to be developed, it would likely take a decade or two for its production to be scaled up significantly and diffused widely amongst consumers. With oil prices already in triple digits and the risks of further spikes looming large, we cannot afford to wait for the appearance of a "magic elixir", as Richard Heinberg has called it (Heinberg 2004).

1.2   Possible Global Consequences

Given the long lead times required to develop alternative transport infrastructure that is less oil dependent (Hirsch et al. 2005), the prospect of an imminent decline in annual oil supplies raises vital questions about the potential impact on societies and their economies, including those of South Africa. Historically, episodes of temporary oil shortages and/or high oil prices, such as in 1973/1974, 1979/1980, and

2007/2008, have been followed in short order by economic recessions in the major industrial countries and in the latter case for the world as a whole (Hamilton 2009). On the upslope of the world oil Hubbert curve, supply was essentially driven by demand: there was always enough oil underground to meet rising consumption (Hall and Ramirez-Pascualli 2012). On the downslope of the Hubbert curve, however, demand will be determined chiefly by (diminishing) available supply. The peaking of global conventional oil production will result most immediately in a supply/demand crunch: growth in desired demand for oil will no longer be met by adequate supplies of conventional oil (with relatively high EROI). Thus future oil prices will be driven higher by (1) stagnant and then declining global supplies of oil and (2) declining average EROI for oil, which pushes up costs of oil production (Heun and de Wit 2011; Murphy and Hall 2011). Hence while the global Hubbert peak does not imply the exhaustion of oil, it does signal the end of the era of cheap oil (Owen et al. 2010). In recent years global oil production has been on an "undulating plateau", which might continue for several more years. Eventually, global production of all oil types will enter a terminal decline phase and by conservative estimates may contract by 2–5 % per annum (Hirsch 2008). Thus on the downslope of the Hubbert curve the world faces the equivalent of an endless sequence of supply-side oil shocks, and prices will have to rise to suppress demand.

While the International Energy Agency has confirmed that the era of cheap oil is over (IEA 2012), its relatively benign price forecasts are contradicted by recent modelling by International Monetary Fund (IMF) researchers and indeed by the data as of mid-late 2013. A paper that examined the interactions of rising demand and geological supply constraints warned that the price of oil could double to $200 per barrel (measured in 2012 dollars) by 2020 (Benes et al. 2012). A second IMF working paper modelled various scenarios for oil prices and their impact on global growth (Kumhof and Muir 2012). In their relatively optimistic "baseline scenario", which assumes (a) that the rate of oil supply growth is constrained to 1 percentage point below the average rate of 1.8 % per annum attained between 1981 and 2005 and (b) that there is a high degree of substitutability of other energy sources for oil, the price of oil nevertheless rises 100 % by 2020 and 200 % after 20 years. If it turns out to be more difficult than expected to find adequate substitutes for oil, or if world oil production begins to decline soon, Kumhof and Muir warn that the oil price could rise to economically devastating levels. Indeed, the historical record shows that international oil price spikes and temporary supply shortages have had serious negative economic impacts in oil-importing nations, resulting *inter alia* in higher rates of price inflation, slower economic growth or recession, deepening poverty and food insecurity, debt crises, and in some cases civil unrest (Hirsch et al. 2005).

While the ramifications of peak oil are likely to be wide ranging, three aspects deserve emphasis. First, as transport costs rise and the reliability of freight transport is undermined by oil supply disruptions, international trade in physical goods (particularly high-bulk, low-value-added items) will decline and supply chains will become shorter. Effectively, higher transport costs will act as a tariff on imported goods (Rubin 2009). As a result the process of globalisation (at least as it pertains to trade in physical goods)—which has depended on cheap, reliable transport—will likely move into reverse (Curtis 2009).

Second, the peaking and decline of world oil production are likely to have a profound impact on global financial markets (Heinberg 2011; Leigh 2008). Fundamentally, the integrity of the world's debt-based financial system is deeply dependent on continuous economic growth. This is because most new money is created as debt, on which interest payments are required. The only way that the interest can be repaid in the future is if more new money is issued, which itself increases the stock of debt. The collateral for this expanding debt is continuous economic growth, which historically has relied on growing supplies of energy. Should economic growth fail for an extended period as a result of energy constraints, the global financial system could implode. In fact as of 2013 the rate of growth of most Western nations had declined to close to zero and debt burdens posed serious risks to financial stability.

Third, as oil production begins to wane after the global peak, international competition for remaining supplies will intensify and place additional strain on geopolitical relations (Klare 2012). Such competition, involving both consumers and producers of oil, could manifest itself in various non-military forms such as bilateral oil deals and trade wars or in the form of armed conflict such as civil wars, terrorism, or inter-country resource wars (Howard 2009). The intensification of these types of conflict would exacerbate world oil supply constraints and shocks in a self-reinforcing feedback loop.

1.3 Mind the Policy Gap

In recent years the nature and implications of global oil depletion have received a growing amount of attention internationally, although opinions as to the potential severity of the consequences of peak oil vary greatly. Thus far, the leaders, governments, or government agencies of only a few countries have publicly acknowledged the threat posed by peak oil—including Sweden, Ireland, France, the United Kingdom, and the United States—but mostly in an oblique and transient manner. More explicit and serious attention has been given to peak oil by the US and German militaries as well as by some industry players. And yet other loud voices—notably from the oil industry and the Organisation of Petroleum Exporting Countries (OPEC)—have kept up a continuous media campaign that tries to convince the public and politicians that "oil's well": that there will be plentiful supplies of oil for decades to come. In addition, despite an increasing global public awareness of the issue of oil depletion, it is receiving limited attention in the academic arena outside of a few mainly energy-specific journals. Most notably, the prospect of an imminent peak and decline in global oil production has largely been ignored within mainstream economic theory and policy. This economic paradigm is at a loss to explain the persistence—or importance—of triple-digit oil prices despite the lingering debt overhang in the industrialised nations and a slowing world economic growth rate.

Table 1.1 Recognition of peak oil in major South African policy documents

Policy document	Year	Recognises peak oil
Medium Term Strategic Framework	2009	No
New Growth Path	2010	No
Industrial Policy Action Plan 2	2011	No
National Framework for Sustainable Development	2008	Yes
National Framework for Local Economic Development	2006	No
Energy Security Master Plan—Liquid Fuels	2007	Obliquely (long term)
Integrated Resource Plan for Electricity 2010	2010	No
National Transport Master Plan	2010	Yes
Comprehensive Rural Development Programme	2009	No
National Spatial Development Perspective	2006	No
National Development Plan	2012	No

Similarly, in South Africa very limited attention has been given to peak oil in academic, public, and policy discourses, despite the country's reliance on imports to meet over two-thirds of its oil demand. As far as existing policy documents are concerned (as of 2013 and reflected in Table 1.1), only the National Transport Master Plan (NATMAP) and the National Framework for Sustainable Development (NFSD) explicitly recognise the phenomenon of peak oil and its potential impacts on South Africa. The Depart of Energy's *Energy Security Master Plan—Liquid Fuels* gives oblique references to future oil scarcity but does not directly or adequately address the looming decline in world crude oil production. Given the crucial role that petroleum plays in the socioeconomic system, a peak oil mitigation strategy should arguably be one of the strategic priorities for all of these major policy documents and should inform policy decisions across all spheres and sectors of government.

1.4 South Africa: A Microcosm of the World

South Africa serves as a useful case study of the implications and mitigation of peak oil because in many ways the country resembles a microcosm of the world at large. It has a modern industrial economy with sophisticated agribusiness, mining, manufacturing, petrochemical, and financial service industries that are on par with those in the Northern industrialised nations. At the same time, almost half of the ethnically diverse population lives in poverty, eking out an existence either in rural areas on small-holder or communal subsistence farms or in urban shanty towns of the type that have mushroomed in the developing world in the past few decades. Thus the South African economy is divided between a formal sector that is closely integrated into the global financial architecture and world trading system and a fairly large and mostly marginalised informal sector.

Table 1.2 Comparison of key indicators between South Africa and the world

Indicator	Year	South Africa	World
GDP per capita (PPP international $)	2011	$10,960	$9,166
Agriculture (% of GDP)	2011	2 %	3 %
Industry (% of GDP)	2011	31 %	26 %
Services (% of GDP)	2011	67 %	71 %
Poverty rate (< $2/day)	2009	31 %	40 %
Poverty rate (< $1.25/day)	2009	14 %	21 %
Income inequality (Gini coefficient)	2008	0.70[a]	0.66[b]
Urbanised population	2010	62 %	52 %
Access to electricity (% of population)	2009	75 %	74 %
Motor vehicles per 1,000 people	2009	162	175
Oil's share of primary energy supply	2009	17 %[c]	33 %[c]
Fossil energy consumption (% of total)	2009	87 %	81 %
Biomass share of primary energy	2009	10.5 %	10 %
Alternative and nuclear energy (% of total)	2009	2.5 %	9 %
Energy use per capita (kg of oil equivalent)	2009	2,921	1,851
Energy use per $1,000 GDP (2005 PPP)	2009	312	181

Source: World Bank, 2013, except where noted below:
[a]Leibbrandt et al. (2010)
[b]Average of figures cited in the Human Development Report 2010 (UNDP 2010: 73)
[c]International Energy Agency (IEA 2013)

Table 1.2 shows a striking degree of similarity between South Africa and the world as a whole across a range of socioeconomic indicators. South Africa's Gross Domestic Product (GDP) per capita of $10,960 in 2011 placed it in the World Bank's upper-middle-income country category and is close to the world average of $9,166 per head. The size of the population was estimated as 53 million in mid-2013 (StatsSA 2013). The division of the economy between agriculture (2 %), industry (31 %), and service sectors (67 %) is almost identical to world averages. The poverty rates in South Africa are only somewhat lower than the global averages, and the extensive degree of income inequality is slighter higher than world inequality as measured by the Gini coefficient. A somewhat higher proportion of South Africa's population is urbanised (62 %) compared to the world urbanisation rate of 52 %, while the proportion of the population with access to electricity is almost identical and the prevalence of motor vehicles relative to population size is very similar. In terms of sources of primary energy, South Africa's reliance on fossil fuels (87 %) versus biomass (10 %) and alternative energy and nuclear power (3 %) is broadly in line with the global averages. The main difference is that South Africa's reliance on oil (17 % of primary energy) and natural gas (3 %) is lower than the world rates, while its dependence on coal (70 %) is higher; but Sasol converts about a quarter of the country's annual coal supply to liquid fuels in its coal-to-liquid plant. South Africa's energy intensity is also somewhat higher than the world average, measured both on a per capita basis and relative to the size of its GDP; this is in large part due to the extensive mineral extraction and processing industries. One of the main socioeconomic differences is that South Africa has a very high official rate of unemployment, estimated at over 25 % in mid-2013 (StatsSA 2013). Environmentally the country faces similar constraints to the world as a whole, including a scarcity of freshwater, declining soil fertility, and vulnerability to climate change.

1.5 Objectives of This Book

This book has two principal aims. The first is to fill the void concerning peak oil in the South African intellectual and policy milieu. The second is to make a contribution to the international literature by providing a country case study that has wider relevance to other developing countries and arguably to the world as a whole. The following five chapters present a sectoral analysis of the implications of peak oil for South Africa and how the likely impacts of declining world oil production can be mitigated. We begin with the "master resource"—energy—before considering the two sectors most heavily dependent on oil, namely, transport and agriculture. Thereafter, the lens is broadened to consider the economy and society at a macro level. The final chapter pulls the thematic analyses together within the conceptual framework of a societal transition towards a more sustainable socioeconomic system and briefly considers the broader significance of this study.

Chapter 2
Energy

The role of oil in South Africa's energy system needs to be placed in the context of overall energy supply and demand balances, i.e. alongside other sources of primary energy and within the final energy consumption mix. Figure 2.1 displays the evolution of South Africa's primary energy supply between 1990 and 2010. None of the relative shares from the various primary energy sources have changed appreciably. Throughout the period, coal has dominated with between 72 and 77 % of primary energy. Oil's share rose notably from a low of 7.1 % in 2002 to over 14 % between 2007 and 2009 but fell back to 11 % in 2010, according to IEA data. Biomass and waste have provided between 9 and 12 % of primary energy, largely consumed by the rural population. The share of gas has remained below 3 %, while nuclear power—generated in Africa's only two reactors—provided 2.3 % in 2010. Renewable electricity, including hydropower, solar, and wind, accounts for less than half a percent of South Africa's energy supply.

Primary energy sources, including crude oil, are converted into energy carriers (e.g. petroleum fuels such as petrol, diesel, and jet fuel), which are then consumed by end users. Coal, petroleum products, electricity, and biofuels and waste have all contributed significant shares of final energy, while use of solar thermal energy remains miniscule (see Fig. 2.2). Over the period 1990–2010, the direct use of coal (i.e. excluding coal that has been transformed to electricity or liquid fuels) has shrunk from over 30 to 23 %, having made way for more efficient energy carriers such as electricity and petroleum. By 2010, petroleum accounted for the largest share of final energy consumption, namely, 30 %.

This chapter explores the implications of peak oil for South Africa's energy system. The first section examines the supply of oil and petroleum products in terms of imports, domestic production, refining and stockpiles, and analyses the demand for petroleum products according to product type and economic sector. The next section briefly considers the likely impacts of oil shocks on the energy system, while the third section presents the main options for substituting alternative energy sources for imported oil and for improving energy efficiency.

J.J. Wakeford, *Preparing for Peak Oil in South Africa: An Integrated Case Study*,   11
SpringerBriefs in Energy, DOI 10.1007/978-1-4614-9518-5_2, © Jeremy J. Wakeford 2013

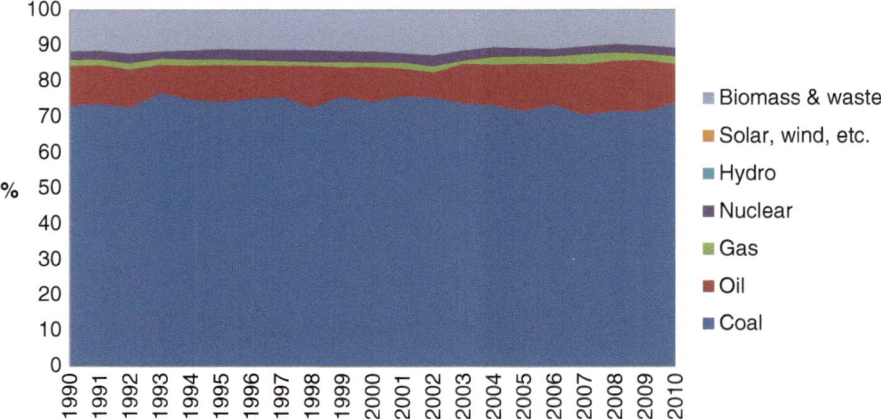

Fig. 2.1 Shares of total primary energy supply by source, 1990–2010. *Source*: IEA (2013)

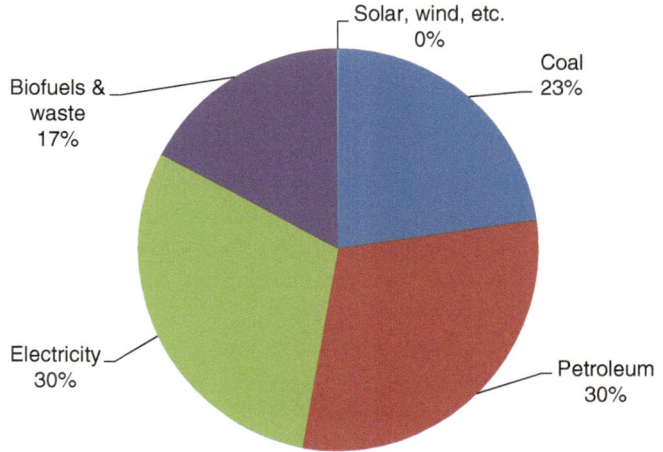

Fig. 2.2 Shares of total final energy consumption by energy carrier, 2010. *Source*: IEA (2013)

2.1   Oil Dependence of the Energy System

2.1.1   Supply of Oil

South Africa has yet to discover any significant crude oil fields. As of the end of 2012 the country's crude oil reserves stood at a meagre 15 million barrels (EIA 2013c) and were likely to be depleted within a few years in the absence of significant new oil field discoveries. The development of South Africa's petrochemical sector is something of

Table 2.1 Historical timeline of South African petroleum industry developments

Year	Development
1950	Sasol (Pty) Limited established by state-owned Industrial Development Corporation (IDC)
1954	Mobil refinery commissioned at Durban
1955	Sasol 1 commissioned at Sasolburg
1963	SAPREF refinery commissioned in Durban—joint venture between Shell and BP
1964	Creation of the Strategic Fuel Fund (SFF) to manage strategic stocks
1965	Refined fuels pipeline commissioned by South African Railways and Harbours (SAR&H) from Durban to Johannesburg via Sasolburg
1965	Soekor established by IDC and government to explore for oil and gas
1966	CALREF refinery commissioned in Cape Town—owned by Caltex (now Chevron)
1967	Crude oil pipeline commissioned by SAR&H from Durban to Johannesburg via Sasolburg
1967	Government began a project to build strategic crude oil stocks at disused coal mines at Ogies
1969	NATREF company formed between Sasol, Total, and National Iranian Oil Company (NIOC)
1971	NATREF refinery commissioned in Sasolburg
1977	UN imposed mandatory crude oil sanctions on South Africa
1977	Central Energy Fund (CEF) established, incorporating SFF
1978	Refined fuel pipeline commissioned by SAR&H from Durban to Johannesburg via Secunda
1979	Sasol purchased NIOC shares of NATREF and became the majority shareholder
1979	Sasol privatised and listed on the Johannesburg Stock Exchange
1980	Sasol 2 commissioned at Secunda
1982	Sasol 3 commissioned at Secunda
1986	Government commenced planning for a new alternative synthetic fuel plant
1989	Mobil sold its SA assets to Gencor, which formed Engen (incorporating Trek)
1992	Mossgas GTL refinery commissioned at Mossel Bay
1993	UN crude oil sanctions lifted
2001	Soekor and Mossgas consolidated to form PetroSA, as a wholly owned subsidiary of the CEF
2000s	Sasol 1 at Sasolburg converted to produce only petrochemical feedstocks
2000s	Strategic oil stockpiles at Ogies sold to Natref
2006	Sasol begins pre-feasibility studies on Mafutha CTL project at Waterberg coal field
2008	PetroSA begins feasibility studies into Mthombo oil refinery at Coega, Eastern Cape
2010	Sasol shelves Mafutha project, citing policy uncertainty regarding the Mthombo refinery and the need for carbon capture and storage
2012	Construction of a new multi-product pipeline from Durban to Johannesburg completed by state-owned Transnet Pipelines
2012	China's Sinopec Group partners with PetroSA to conduct front-end engineering design (FEED) for Project Mthombo refinery

Source: Adapted from Rustomjee et al. (2007)

an anomaly in the global context—a consequence of its lack of indigenous crude oil and a legacy from its isolation in the apartheid era (see Table 2.1 for a historical overview of major developments). After coming to power in 1948, and mindful of the fate of oil-poor Nazi Germany, the Nationalist Government was keenly aware of the critical importance of energy security—especially in light of the country's energy-intensive, mineral-dominated economy. This led in 1950 to the creation by the state of the Suid-Afrikaanse Steenkool-, Olie- en Gasmaatskappy (South African Coal, Oil and Gas Company), abbreviated "Sasol". Sasol began producing coal-to-liquid (CTL)

Fig. 2.3 Map of South Africa showing location of key energy sites. *Source*: Adapted from EIA (2013c)

synthetic fuels at Sasolburg in 1955, and, following the imposition of United Nations sanctions on the Republic in 1977, a newly privatised Sasol Limited expanded its CTL capacity to approximately 160,000 barrels per day (bpd) in the early 1980s at Secunda (in what is now Mpumalanga Province). Subsequently, the apartheid state commissioned the world's first commercial gas-to-liquid (GTL) plant at Mossel Bay on the country's southern coast (Fig. 2.3), which began production in 1993 with a capacity of 45,000 bpd. Mossgas was later merged with the state's oil exploration company, Soekor, to form the national oil company PetroSA. Today Sasol is one of the country's largest and most profitable companies, and the diverse group has interests in numerous countries around the globe, including leading GTL developments in Qatar and Nigeria. Thus as a result of its isolationist past, South Africa boasts two of the world's leading developers of synthetic petroleum products.

In 2011 South Africa is estimated to have consumed 610,000 bpd of petroleum products, of which approximately 430,000 bpd (71 %) was met by imported crude oil and refined products (EIA 2013c). The remaining domestic production of 180,000

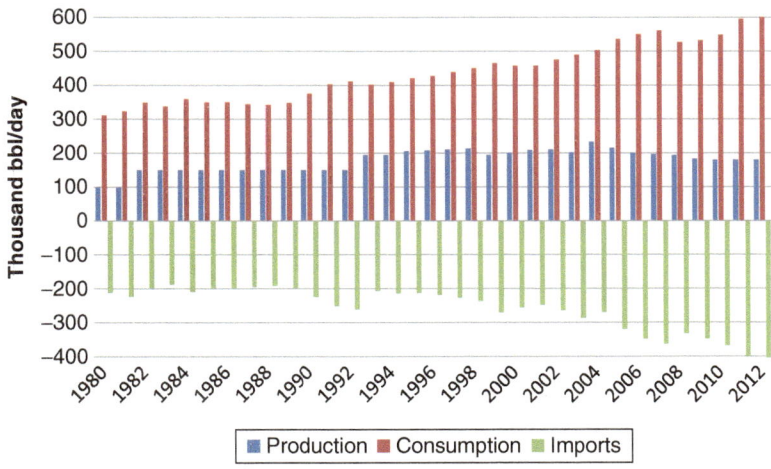

Fig. 2.4 South African oil product production, consumption, and imports, 1980–2012. *Source*: EIA (2013a)

bdp was derived from Sasol's CTL synthetic fuels (160,000 bpd or 26 % of the total) and PetroSA's production of GTL synthetic fuels (14,000 bpd) plus a very small amount of domestic crude oil and condensate (6,000 bpd), which together contributed just 3 % of total petroleum supply. Figure 2.4 displays South Africa's total annual production, consumption, and imports of oil products (crude oil plus refined petroleum products). Domestic production has remained relatively constant in a range between 180,000 and 200,000 bpd since 1993, while consumption has followed a rising trend albeit with some cyclical movements related to the rate of economic growth. Oil imports have therefore been on a gradually rising trend since 1993.

In 2011, South Africa relied mostly on OPEC nations for its oil imports, notably Iran (27 %), Saudi Arabia (27 %), Nigeria (20 %), Angola (18 %), and Oman (7 %) (see Fig. 2.5). However, reliance on Iranian crude oil imports was curtailed to a significant extent in 2012 and 2013 under pressure from the United States and the European Union, which placed sanctions on the Iranian oil industry. South Africa sourced more oil from its African neighbours and Saudi Arabia to replace the Iranian oil.

South Africa has a long history of oil refining, with the first refinery built by Mobil (currently owned by Engen) in Durban in 1954, followed by another constructed by Shell and BP (Sapref) in 1963, and a third built in Cape Town by Caltex in 1966 (now owned by Chevron). Since most of the refineries were located on the coast but a major share of fuel consumption took place in the industrial heartland surrounding Johannesburg (see Fig. 2.3), a refined fuels pipeline from Durban to Johannesburg via Sasolburg (in the Free State Province) was commissioned by the state-owned South African Railways and Harbours (SAR&H) company in 1965. This was followed 2 years later by a crude oil pipeline to feed a new inland Natref refinery at Sasolburg, which was a joint venture between Sasol, Total, and the National Iranian Oil Company (NIOC). Today South Africa has the second largest

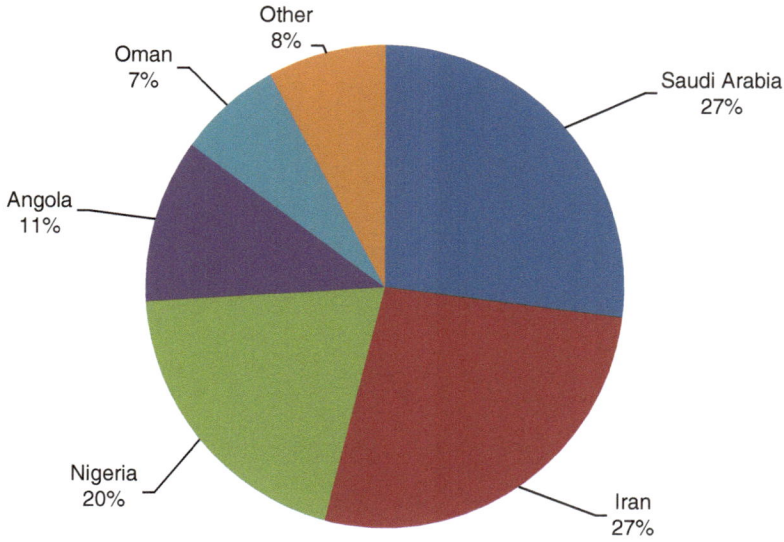

Fig. 2.5 South Africa's sources of crude oil imports, 2011. *Source*: EIA (2013c)

oil refining capacity in Africa, after Egypt. Total refining capacity in 2012 amounted to a nominal 703,000 bpd, of which 72 % comprised crude oil refining capacity with the balance of 28 % being synthetic fuel refining capacity (see Table 2.2). For many years South Africa has exported refined petroleum products to neighbouring countries in Southern Africa. From 2006 demand for refined fuels in the region (including South Africa) outstripped domestic refining capacity so that increasing amounts of refined fuels had to be imported. Petroleum fuels in South Africa are distributed from the refineries to approximately 200 depots, to 4,600 retail service stations, and directly to about 100,000 consumers, most of whom are farmers (SAPIA 2013). The existing crude oil refineries were constructed many decades ago and are due to be upgraded over the coming years to comply with cleaner fuel standards.

In the apartheid era the government created a Strategic Fuel Fund (SFF) to manage strategic oil stocks (see Table 2.1). A project was initiated in the late 1960s to convert abandoned coal mine shafts to oil storage facilities, but these stocks were sold to Natref in the early 2000s when oil prices were comparatively low. Currently South Africa maintains a strategic petroleum reserve at Saldanha Bay in the Western Cape Province. The facility has a maximum capacity of 45 million barrels, which currently translates into about 110 days' worth of oil imports (DME 2007a). Information on the actual volume of oil in storage is not publicly available. In December 2005 South African oil refineries underwent modifications in order to comply with cleaner fuel regulations, and shortages of refined product emerged in certain areas, which brought about economic losses and inconveniences. In view of this, the Department of Minerals and Energy recommended that the oil industry be required to maintain 28 days' worth of commercial petroleum product stocks (DME 2007a).

Table 2.2 Domestic crude oil and synthetic fuel refining capacity, 2012

Refinery	Barrels/day	Location	Company
Natref	108,000	Sasolburg	Sasol/Total
Sapref	180,000	Durban	BP/Shell
Enref	120,000	Durban	Engen
Chevref	100,000	Cape Town	Chevron
Total crude oil refining	508,000		
Secunda	150,000	Secunda	Sasol
Mossgas	45,000	Mossel Bay	PetroSA
Total synthetic fuel refining	195,000		
Total	703,000		

Source: SAPIA (2013)

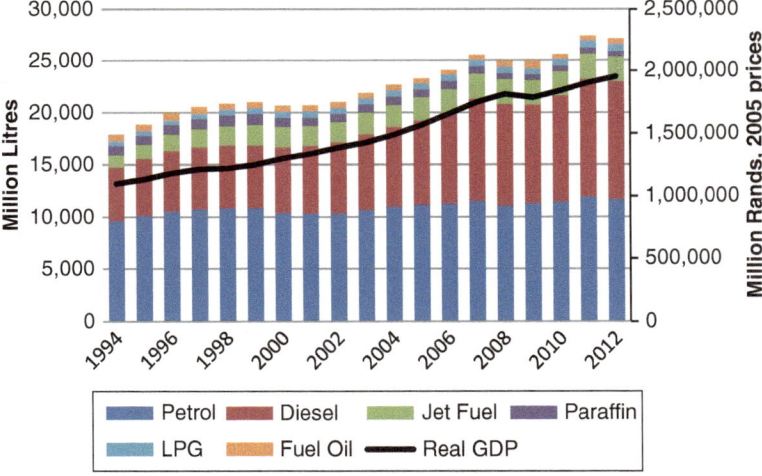

Fig. 2.6 Annual total petroleum product sales and real GDP, 1994–2012. *Source*: SAPIA (2013); SARB (2013)

2.1.2   Demand for Oil

Total annual sales of petroleum products grew largely in line with the economy (real GDP) in the period 1994–2012 (see Fig. 2.6). Petrol (gasoline) and diesel together make up more than 80 % of petroleum product sales. Liquefied petroleum gas (LPG) is used mainly for household use for cooking and heating, while paraffin (kerosene) remains an important fuel for lighting and cooking amongst poor households. The relative shares of petroleum products partly reflect demand and partly the proportions of a barrel of oil that can be refined (or "cracked") into the various products.

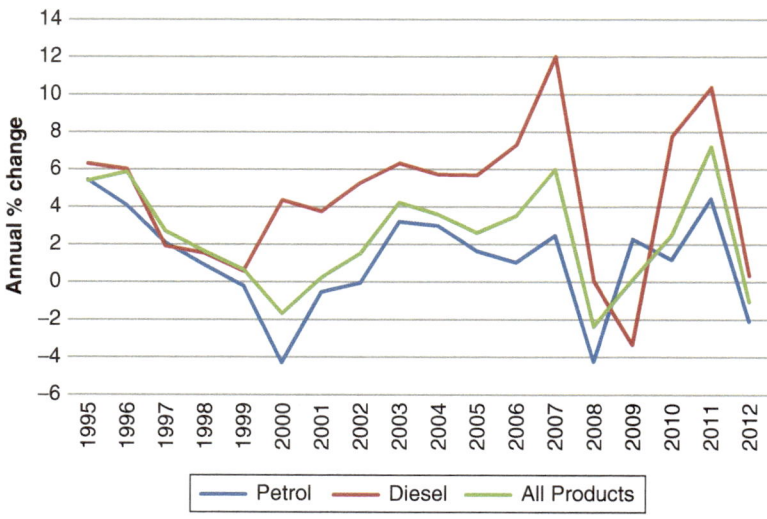

Fig. 2.7 Annual growth in petroleum product sales, 1995–2012. *Source*: SAPIA (2013)

The average growth rate for sales of all liquid petroleum fuels was 2.8 % for the period 1995–2007 (see Fig. 2.7). In that period the average annual growth rate for diesel was 5.1 % and for petrol, 1.4 %. However, these growth rates fell steeply during 2008 as a result of sharply rising fuel prices (crude oil traded at nearly $100 per barrel on average for the year) as well as tighter economic conditions, i.e. rising costs of living and higher interest rates. The recession in 2009 dampened demand for diesel (consumption fell by 3.3 %), although petrol demand grew by 2.3 % thanks to the drop in oil and petrol prices.

The transport sector accounted for the lion's share (69.8 %) of total oil product demand in 2010, while the shares consumed by other sectors were rather small: agriculture (4.4 %), industry (4.3 %), commercial and public services (3.8 %), residential (2.7 %), and non-energy uses (14.9 %) (IEA 2013). The dependence on petroleum, as measured by the shares of petroleum products in total final energy consumption, for the main economic sectors in 2010 is illustrated in Fig. 2.8. The transport sector is almost entirely dependent on petroleum (i.e. almost 98 % of the sector's total energy is derived from petroleum fuels), while agriculture relies on oil products for 59 % of its energy supply. Industry uses mainly electricity and coal and relies on petroleum for less than 4 % of its energy supply. The residential sector relies very little on petroleum for household energy use (3.4 %), while petroleum dependence is somewhat higher in the commercial and public services sector (16.3 %).

Figure 2.9 shows the per capita consumption of petroleum products for the period 1994–2012. There has been a very slightly increasing trend over the period, albeit with some undulations. Per capita demand for diesel has grown strongly and has almost caught up with that of petrol, which has been declining slightly since peaking in 1997.

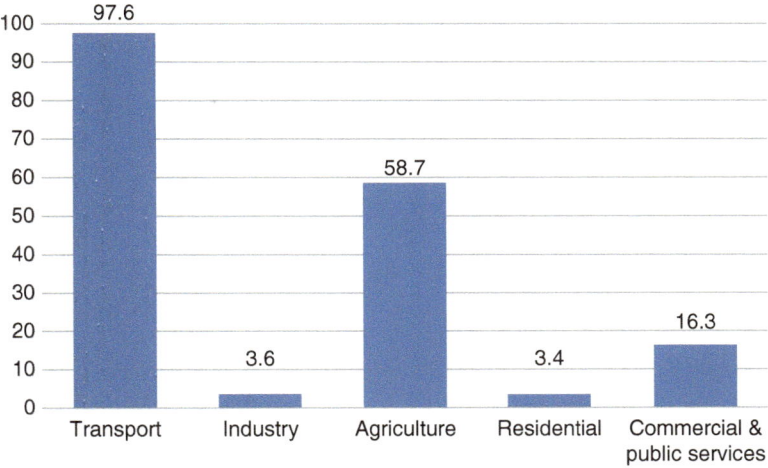

Fig. 2.8 Share of petroleum products in final energy consumption by sector, 2010. *Source*: IEA (2013)

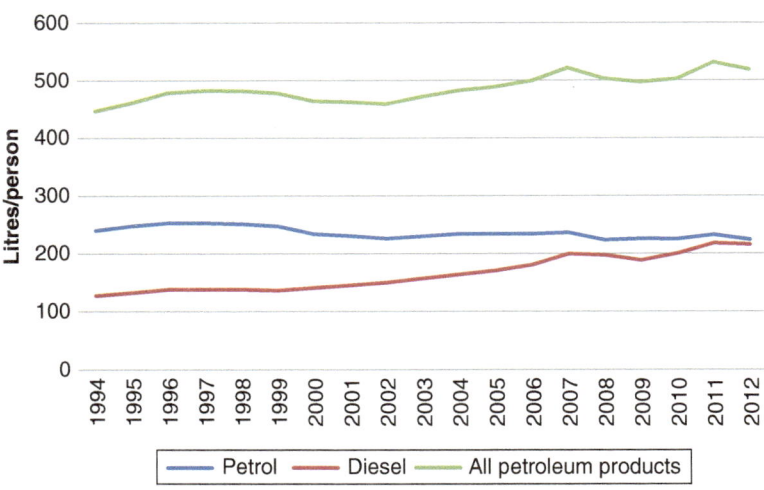

Fig. 2.9 Petroleum consumption per capita in South Africa, 1994–2012. *Source*: SAPIA (2013); StatsSA (2013)

2.2   Likely Impact of Oil Shocks

The peaking of world oil production will most likely result in a rising oil price trend, albeit with greater volatility, and the emergence at some point of physical supply shortages. Rising oil prices will gradually dampen demand and result in less petroleum energy being consumed in the country, especially in the longer term.

The prices of other energy sources, especially those that are to some extent or other substitutable for oil—such as coal and gas—are likely to rise along with the oil price. These price rises will in turn put upward pressure on the price of electricity, since coal is the feedstock for about 90 % of national power generation. Furthermore, because approximately one-third of the coal feeding state-owned utility Eskom's coal-fired power stations is transported by truck, the costs of this feedstock will rise as diesel prices rise. Higher prices of refined diesel fuel will also raise Eskom's costs of running open-cycle gas turbines (OCGTs), which are used to meet peak electricity demand (and despite their name run on diesel, not gas). The costs of buying or manufacturing, transporting, and installing alternative energy infrastructure, including wind turbines and solar panels, will also increase to some extent (all else being equal) as a result of rising petroleum fuel costs, reflecting their dependence on an economic infrastructure that is itself dependent on oil. Overall, there will be added upward pressure on electricity prices, in addition to the pressure imposed by funding requirements for Eskom's new build programme. The rising cost of fossil fuel energy will make renewable energy (RE) sources relatively more competitive and is likely to stimulate investment in this sector. Increased production of RE technologies could deliver economies of scale and learning and hence reduce their prices, setting off a positive feedback loop. Thus over the longer term, one can expect a process of (partial) substitution of renewable energy for oil and coal. If economic conditions are deteriorating (as discussed in Chap. 5), however, the expansion of RE might not be rapid enough to offset declining consumption of fossil fuels, resulting in diminishing total energy consumption. In addition most alternative energy options are not direct substitutes for liquid petroleum fuels.

Acute physical shortages of oil products, which could arise from time to time owing to global supply interruptions, could have even more serious consequences than gradually rising (or volatile) energy costs. Most immediately, Eskom's demand for diesel fuel to run its OCGT peaking power plants will have to compete with transport, agriculture, and other demand sectors for scarce diesel supplies. Perhaps most significantly, a sudden interruption of liquid fuel supplies could disrupt the flow of coal to power stations and thereby seriously compromise Eskom's ability to maintain sufficient power generation to keep the national electricity grid stable. Although not caused by liquid fuel shortages, a similar situation arose in early 2008 when problems in the procurement and transportation of coal resulted in insufficient stockpiles at some power stations, contributing to the electricity crisis which involved extensive blackouts and load shedding. Power outages would in turn hamper the refining of petroleum fuels and their distribution through pipelines and at retail outlets, thus setting in motion a self-reinforcing feedback loop with very adverse consequences.

2.3 Weaning Off Oil: Energy Substitutes and Conservation

In terms of mitigating the impacts of global oil depletion, the primary goal for the energy sector is to reduce reliance on imported oil. This should involve a combination of curtailment of demand for oil products and a shift to more sustainable energy sources. The government's main approach to liquid fuel supply is summarised in the *Energy Security Master Plan—Liquid Fuels* (DME 2007a). This document recommended a set of mostly infrastructural short- to medium-term strategies and longer term strategies which essentially amounted to the development of modelling capacity and policy formulation. Specific recommendations made by the Department of Minerals and Energy included the procurement by PetroSA of 30 % of South Africa's crude oil supply, apparently in order to reduce the risks of reliance on a few sources of oil imports; the promotion of local liquid fuel production; mandating energy companies to ensure adequate commercial fuel inventories; and the cross-sectoral promotion of energy efficiency, noting in particular the risks of over-reliance on oil by the transport sector.

The infrastructural strategies emphasised the need for adequate quantities of refined fuels to be made available to meet rising demand, especially in the economic heartland of Gauteng Province. In view of this, two major petroleum infrastructure projects were initiated. The first is Transnet Pipeline's new multi-product fuel pipeline connecting the port city of Durban with the inland Gauteng Province, which was commissioned early in 2012. The second is PetroSA's proposal for a new oil refinery, dubbed "Project Mthombo", to be located at the port of Coega near the city of Port Elizabeth in the Eastern Cape Province (see map in Fig. 2.3). Mthombo was originally slated to have a capacity of 400,000 bdp, which would make it the largest refinery in Africa, and early projections put the cost at around $11 billion (approximately R110 billion). PetroSA is partnering with Chinese firm Sinopec to investigate the front-end engineering and design for the refinery. As of this writing, however, the government has yet to approve the project.

Both the new pipeline and the proposed refinery assume that increasing quantities of crude oil will be available globally and affordable over their lifespans (of at least 50 years). Like the *Liquid Fuels Master Plan*, these projects take no account of the future limitations on crude oil imports that will be imposed sooner or later by declining world oil exports and rising oil prices. To address this vital policy deficiency, this section explores the alternative sources of energy supply that could substitute for imported crude oil and refined petroleum fuels. I also discuss demand-side management, in terms of energy efficiency and conservation initiatives, and motivate for South Africa to sign an international Oil Depletion Protocol to guide a co-ordinated reduction in oil consumption.

2.3.1 Developing Domestic Energy Alternatives

South Africa's state-owned oil company PetroSA produces approximately 1,800 bpd of crude oil from its Oribi and Oryx fields off the southern tip of the country, about 0.3 % of the nation's oil use (PetroSA 2012). Although oil exploration is continuing off the western and southern coasts, no new oil discoveries have been announced as of this writing (November 2013), and therefore there is no expectation of a notable increase in domestic crude production in the foreseeable future. Hence, the realistic prospects for domestic liquid fuel production rest on CTL, GTL, and biofuels, which are considered in turn below. Aside from these sources, biogas can potentially substitute for liquefied petroleum gas, while electricity represents a general—although imperfect—energy replacement for petroleum fuels.

2.3.1.1 Coal to Liquids

Sasol currently supplies approximately 26 % of South Africa's annual liquid fuel demand from its CTL plant at Secunda in Mpumalanga Province (see map in Fig. 2.3). The major advantage of CTL is that it is a reliable technology with a proven track record, which produces synthetic petroleum fuels (synfuels), including petrol, diesel, and jet fuel, that are usable in existing transport infrastructure. Expanding domestic CTL production would therefore reduce South Africa's dependency on oil imports and save foreign exchange. Although information on Sasol's CTL production costs is not publically available, it may be assumed based on the company's profitability that they are less than the cost of refined fuel imports. In March 2010 Sasol's board approved the first phase of a project to expand the synfuels and electricity generation capacity of the Secunda plant by approximately 3.2 %, using natural gas imported from Mozambique as feedstock (Sasol 2010). The Secunda expansion is due to come on stream in 2014.

Sasol has also investigated the viability of a new CTL plant to be located at the Waterberg coal field in Limpopo Province (Sasol 2010: 20) (see map in Fig. 2.3). Named Project Mafutha, the proposed plant had a capacity of 80,000 barrels of liquid fuels per day, about half of Sasol's existing synfuel production volume. Sasol has indicated that it would not be the sole investor in such a large-scale project, which was estimated to cost in the region of R160 billion (around $16 billion), and the company sought financial support from government (Njobeni 2010). In 2008 Sasol signed a Memorandum of Understanding with the Industrial Development Corporation for a planned investment in the project, and the company also held investment talks with the departments of Trade and Industry and Minerals and Energy. According to Sasol, Project Mafutha would likely take up to 10 years to complete. If both Project Mafutha and the Secunda extension materialised, Sasol's synfuels would meet about 40 % of the country's 2010 liquid fuel demand.

Construction of a new CTL plant faces several risks and would entail costs other than purely financial costs. First, such a project would be viable only if sufficient coal feedstock could be secured for the lifetime of the project. While the Waterberg coal field is relatively underutilised, South Africa's remaining coal reserves are the subject of much contention. The official government figure for reserves was revised downward greatly from over 50 gigatonnes (Gt) to under 30 Gt in 2007 (GCIS 2007). However, recent research casts doubt on even this latter figure. Professor David Rutledge estimates that remaining recoverable coal reserves in Southern Africa (the vast majority of which are in South Africa) may be as low as 10 Gt (Rutledge 2011). Based on historical production data and the "Hubbert linearisation" method, local geologist Chris Hartnady forecasts a peak in domestic coal production at about 284 million tonnes per annum (mtpa) in 2020, up slightly from the 2012 level of production of 260 million tonnes (Hartnady 2010). On the other hand,

Eskom's demand for coal for electricity generation is set to rise by approximately 30 mtpa (to feed its new Medupi and Kusile power plants) to a peak of around 155 mtpa in 2021, thereafter declining as old power plants are decommissioned (Eberhard 2011). Meanwhile, the coal industry has plans to increase exports from about 65 mt in 2010 to over 90 mt by 2020. The proposed Mafutha CTL plant would require approximately 25 million tonnes of additional coal per annum. If the conservative coal production forecasts noted above turn out to be accurate, then coal production in the country as a whole will not be able to rise sufficiently to meet projected growth in demand by Eskom, other domestic users, exports, and a new CTL plant. Trade-offs amongst these competing uses of coal would have to be made at some point, and domestic coal prices would likely rise considerably. Under these circumstances, it might make more sense for the Waterberg coal to be used to maintain electricity production from existing power plants rather than to feed a costly new CTL plant. Some of this electricity could be used to power transportation (e.g. electric trains and road vehicles), as discussed in the next chapter.

The second risk to building a new CTL plant is that, even if sufficient feedstock were procured, the energy return on investment (EROI) for CTL could be rather low. While the energy content of a tonne of coal (22.75 mBtu) is more than four times the energy contained in a barrel of oil (5.45 mBtu), the CTL process produces between 1 and 1.4 barrels of synfuel per tonne of coal (Höök and Aleklett 2010). This demonstrates that there is a significant "energy price" to pay for converting coal into liquid fuels, which is compensated by the higher prices attracted by liquid fuels on the market due to their convenience. While there are no estimates for the EROI of coal or CTL in South Africa, international estimates for the EROI of coal range from between 60:1 and 80:1 in the United States to around 21:1 in China (Lambert et al. 2012). The EROI for coal mining can be expected to decline over the long term as the quality of ore grade diminishes, hence raising production costs. An experienced local geologist has also raised questions about the quality and accessibility of the Waterberg coal deposits (Hartnady 2010).

A third risk to expansion of CTL capacity is posed by the potentially high environmental costs in the form of water and air pollution, including additional greenhouse gas (GHG) emissions, which contribute to climate change. In view of South Africa's climate mitigation commitments under the Copenhagen Accord of 2009, Sasol may be required to install carbon capture and storage (CCS) technology at a new CTL plant, which would raise its costs considerably. Costs of CTL fuels would also rise following the announcement by the Minister of Finance in February 2013 that a carbon tax would be phased in from 2015. In addition, CTL facilities require large quantities of water, which is an increasingly scarce resource in Southern Africa in general and in the Waterberg area in particular (Hartnady 2010). Finally, the pollution resulting from coal mining and combustion can also have negative impacts on human health, such as respiratory diseases (Spalding-Fecher and Matibe 2003).

In view of these risks, and also because of the unaffordability of the projected financial costs, Sasol put Project Mafutha on hold in late 2010. Given the substantial lead times required for new investments of this scale, it is probably safe to assume that no new CTL plant will be built in South Africa for the remainder of this decade at least.

2.3.1.2 Gas to Liquids

PetroSA produces liquid fuels using natural gas feedstock at its GTL refinery at Mossel Bay on the southern coast (see map in Fig. 2.3). Maximum production capacity is 45,000 bpd of synfuels, although in recent years actual production has been curtailed to less than half of this amount, owing to maintenance issues and gas feedstock supply constraints (PetroSA 2012). The existing gas fields in the Bredasdorp basin, including the newly authorised F-O field (dubbed Project *Ikwhezi)*, are expected to last until at least 2018. The company states, "Further development of other gas prospects near the F-O field could potentially help to sustain the life of the Mossel Bay refinery until 2025". PetroSA is also conducting exploration activities off the country's west coast but as of this writing had not announced any discoveries.

There are at least three other potential sources of natural gas that could supply feedstock to the Mossel Bay GTL refinery or possibly even a new GTL plant (which could be built by either Sasol or PetroSA): imported gas, shale gas, and underground coal gasification. In recent years there have been very substantial discoveries of conventional natural gas offshore of Namibia and Mozambique, which led the Department of Energy and PetroSA to explore the feasibility of importing liquefied natural gas (LNG). LNG has to be transported in special tanker ships and then regasified before it can be used onshore, which requires costly new infrastructure. In 2010 PetroSA's management decided against the LNG option and chose the local development project *Ikwhezi* instead (PetroSA 2010). However in July 2013 PetroSA announced that it was planning to build a floating LNG terminal near its Mossel Bay refinery and appointed an Australian company to conduct a front-end engineering design study with a view to making a final investment decision in late 2014 (Burkhardt 2013). Although LNG prices have in the past been quite closely correlated to oil prices, the development of shale gas in North America over the past few years has lowered gas prices in that region and also softened world LNG prices. However if South Africa does pursue the LNG option, it will have to compete on the global LNG market with the likes of China, Japan, India, and South Korea.

Another potential source of feedstock for GTL plants, albeit highly contentious, is shale gas. In April 2011 the South African Cabinet placed a moratorium on shale gas exploration and appointed an interdepartmental task team to investigate the economic, social, and environmental implications of shale gas development. The Working Group on Hydraulic Fracturing delivered its report in July 2012 (DMR 2012), and the report was subsequently endorsed by the Cabinet. A study commissioned for the US Energy Information Administration (EIA 2011) indicated that South Africa may have potential for shale gas deposits in the Karoo Basin (which underlies a large central portion of the country) amounting to 485 trillion cubic feet (Tcf) of technically recoverable resources. This estimate was reduced to 390 Tcf in a follow-up report (EIA 2013b). However, the Working Group stated that "owing to the limited amount of available data in the area, it is impossible to quantify the resource accurately, other than to say that it is potentially very large". Experience from the United States and other countries suggests that the commercially viable portion of shale gas resources is likely to be much smaller than the technically recoverable resource (Berman 2010; Hughes 2011; Hughes 2013). Furthermore, serious concerns have been raised about potentially negative social and environmental side effects related to the contamination of water and air pollution (Howarth et al. 2011; Hughes 2011). Of particular concern is the limited availability of and possible contamination of freshwater, which is a very scarce resource in the Karoo area.

In September 2012 the Cabinet endorsed the lifting of the 18-month moratorium on shale gas exploration upon the recommendations of the task team. However, only "normal" exploration methods, and not hydraulic fracturing, would be allowed for an initial 6–12-month period while the regulatory framework was augmented. The companies that have been awarded exploration licences still have to complete Environmental Impact Assessments before any exploratory drilling can take place. According to the Working Group report (DMR 2012: 29), "It may take 10 or more years for a successful project to progress from the issuing of an exploration right, through the drilling of a discovery well, the drilling of a number of appraisal wells, the development of an economic feasibility plan, the application for and issuing of a production right, the drilling of production wells and the installation of the pipeline infrastructure before gas is delivered to the end user". As of this writing, therefore, the potential of shale gas to contribute to the energy supply in South Africa remains uncertain, and it seems unlikely to play a meaningful role this decade but could potentially have a major impact on domestic energy markets after 2020. If a commercially recoverable resource of, say, 30 Tcf were established, this could potentially sustain PetroSA's current operations and possibly provide feedstock for new GTL production.

A third source of feedstock for GTL could potentially come from a process called underground coal gasification (UCG), a process whereby coal is ignited in situ underground, fed through a borehole by air or oxygen to yield a synthetic gas (syngas). The syngas can be used for electricity generation, for the production of synthetic liquid fuels or for industrial uses. In addition to this flexibility, several other advantages are claimed for UCG (Eskom 2010; Shafirovich and Varma 2009). First, otherwise uneconomical resources can be utilised; Eskom estimates that an additional 45 billion tons of coal could be exploited through UCG, over and above existing proved reserves. Second, capital investment costs are lower than for conventional coal plants. Third, there are no costs incurred for transporting coal. Fourth, there is no need for traditional mining, and therefore associated health and safety risks for miners are reduced. Fifth, indications from a pilot UCG project in Australia indicate that the process has a much lower environmental impact (in terms of groundwater contamination, land degradation and subsidence, and greenhouse gas emissions) when compared to conventional coal mining. Eskom has a small pilot UCG plant in operation at its Majuba power station in Mpumalanga and began commercial co-firing of gas and coal in October 2010. Eskom is optimistic that the costs will compare favourably with those of conventional coal mining and power generation.

Nonetheless, there are several disadvantages and risks attached to UCG. First, although UCG might produce a smaller volume of GHGs per unit of energy than conventional coal, there are still considerable emissions to deal with. Second, there are concerns about possible underground water contamination and land subsidence (Shafirovich and Varma 2009). UCG has yet to be proven on a commercial scale and thus is a highly uncertain potential contributor to gas supplies in SA. In any event, since the coal fields are located in the northern areas of the country while PetroSA's GTL refinery is in the southern Cape, costly pipeline infrastructure or a new GTL plant would be required to convert coal gas into liquid fuels.

In conclusion, it is reasonably assured that PetroSA will continue to produce GTL from its Mossel Bay refinery until at least 2018 using gas from the southern Cape offshore fields. Beyond that, there are various possibilities for expanding GTL production from domestically produced gas (if new conventional fields are found or if shale gas is found and developed) or from imported gas. However, each of these options would require costly infrastructure investments and could have seriously detrimental environmental side effects.

2.3.1.3 Biofuels

In December 2007 the South African Government approved a *Biofuels Industrial Strategy* (DME 2007b). Citing food security concerns, the Strategy excluded maize as a feedstock for ethanol, advocating instead grain sorghum, sugar cane, and sugar beet. The Strategy also proposed that biodiesel be produced from soya beans, canola, and sunflower oil. The target for biofuel penetration was set as 2 % of liquid road fuels by 2013, in an initial 5-year pilot phase. In August 2012 the Department of Energy gazetted regulations pertaining to the Mandatory Blending of Biofuel with Petrol and Diesel in South Africa, although the implementation date is still to be determined by the Minister of Energy. The regulations stipulate that bioethanol must comprise between 2 and 10 % of petrol on a volumetric basis, while diesel should have a minimum concentration of 5 % of biodiesel volumes.

Obstacles to the development of biofuels in South Africa thus far have included the following factors: low levels of awareness about the opportunities inherent in biofuels; technical challenges; food insecurity concerns; water scarcity; difficulties accessing financing; human capacity constraints; and an uncertain policy and regulatory environment (Amigun et al. 2008; Chakauya et al. 2009). Although large-scale production of biofuels may now become viable under the new regulations, the constraints imposed by water and arable land scarcity suggest that it is unlikely that biofuels will make a significant contribution to national liquid fuel supplies beyond what is envisaged in the blending regulations, i.e. approximately 5 % of current liquid fuel demand (about 30,000 bpd). Perhaps more importantly, the real contribution of biofuels will in all likelihood be severely limited by low EROI ratios, which have been estimated at averaging around 0.9:1 for biodiesel and 1.3:1 for ethanol (Lambert et al. 2012).

For the longer term, there may be the possibility for the so-called second-generation biofuels, such as cellulosic ethanol, which utilises non-food crops, agricultural waste, and wood chips as feedstock (Woodson and Jablonowski 2008), and biodiesel produced from algae (Rhodes 2009). The main problem with cellulosic ethanol is that cellulose is much, much harder to break down than starch, which lowers the net energy yielded by the conversion process. Furthermore, there is no ecological "free lunch": for arable land to remain fertile, a significant proportion of the nutrients contained in the biomass must be returned to the soil—the more so when synthetic fertilisers become relatively scarcer and more costly. The purported benefits of microalgae are rapid growth rates, high oil content, and high yields and the fact that it can be grown using saline water or wastewater. But despite hundreds of millions of dollars being spent on research over the past decade, no significant volumes have been produced to date and costs are very high (Hall and Benemann 2011). Commercialisation of microalgae will require improved algae strains and innovations that lower harvesting and oil-extraction costs and boost co-product output. Thus second-generation biofuel technologies are still in the research and development stage, and it will likely take a decade or more before they can be successfully commercialised.

2.3.1.4 Biogas

Biogas, which is generated from the anaerobic fermentation of organic material, is a substitute for LPG. It has several advantages: (1) it can be produced from organic waste matter and therefore control pollution, including GHG emissions; (2) useful by-products include fertiliser and water; (3) production technology is simple and efficient at both large and small scales, in rural and urban settings; (4) it alleviates pressure on wood resources, deforestation, and related environmental impacts; and (5) biogas systems can be constructed and operated locally (Amigun et al. 2008: 701–2). Capital costs represent the largest component of biogas costs, while operation and maintenance costs are relatively low and the feedstock is often free as it consists of various waste materials. The biogas can be used for heating and cooking, or it can be converted into electricity.

Nevertheless, biogas does have several drawbacks, such as the energy losses that occur in the conversion from biomass to gas and then to heat or power, and the need for the digester to be warm or insulated. More importantly, biogas is a diffuse energy source relative to fossil fuels, which means that if production is to occur at a significant scale then feedstock will have to be transported long distances. It has been estimated in the South African context that transport costs are the second largest component of biogas manufacturing costs, and therefore decentralised plant location close to feedstock sources (and final consumption) is important (Nolte 2007). Biogas therefore presents a good opportunity for sustainable energy supply but on a relatively small, local scale. It has been estimated that 300,000 (mainly rural) households could utilise biogas digesters in South Africa (Trollip and Marquard 2010).

2.3.1.5 Electricity

Electricity can in principle replace petroleum-based transport fuels, although this would require a costly and time-consuming replacement of the current petroleum-based vehicle fleet with electrified transport infrastructure (various options for which are discussed in Chap. 3). Electricity can also substitute for paraffin (which is used chiefly by low-income households for lighting and cooking) and LPG (also used for cooking). However, these new uses of electricity would need to compete with other demand sectors such as industry, commerce, agriculture, and residential. By 2007–2008, demand for electricity had already outstripped Eskom's supply capacity, resulting in a near-collapse of the grid and subsequent power rationing and demand restrictions. Eskom has warned that its generation capacity will be severely constrained until at least 2017, when the first unit of the second new base-load coal-fired power station (Kusile) is due to be commissioned (DoE 2011).

In 2011 the South African Department of Energy published an "Integrated Electricity Resource Plan for South Africa—2010 to 2030" (IRP2010), which projected future electricity demand on the basis of an assumed economic growth rate of 4.6 % per annum and spelled out how generating capacity would increase to meet this demand (DoE 2011). Of the 45 gigawatts (GW) of new capacity envisaged in the IRP2010, 42 % is renewables, 23 % nuclear, and 15 % coal (which excludes 10.1 GW of new coal-fired capacity that was already "committed"). However, the share of electricity actually *generated* from renewable sources was forecast to be just 9 % in 2030 (excluding 5 % for hydro), as a result of the lower load factors for solar and wind power compared to other sources. The IRP2010 also includes 7.3 GW of capacity from OCGTs, fed by diesel. Interestingly, the Department of Energy partially justifies its decision to build 9.6 GW of new nuclear generation capacity on the basis that it "should provide acceptable assurance of security of supply in the event of a peak oil-type increase in fuel prices", which would likely make operation of the OCGTs prohibitively expensive (DoE 2011: 14). As of August 2013, however, the government had not yet given the green light for the nuclear build programme, as it investigated the massive costs and the possibility of tapping natural gas from Namibia and Mozambique instead. Even if the new nuclear build programme does proceed, it will not produce power before about 2024 at the earliest. If exploration for shale gas in the Karoo Basin yields substantial commercial reserves, this could be a potential game changer in the power sector—although again not before the early 2020s and with potentially serious environmental and health consequences.

The Department of Energy's projections for electricity demand growth are probably over-optimistic in light of sluggish economic growth in an adverse global environment, together with the doubling of electricity tariffs that occurred between 2008 and 2013 and the recent decision by the regulator to allow further increases of 8 % a year for the next 5 years. On the other hand, the IRP2010 does not advocate replacing liquid fuels with electricity for transport (the case for which is presented in Chap. 3). The country's further development is therefore likely to require substantial increases in power capacity. Leaving gas aside, the future arguably lies in renewable energy sources. South Africa is a water-scarce country and has largely exploited its large-scale hydropower potential, while biomass cogeneration and landfill waste will make relatively small contributions to the power mix in the foreseeable future.

The best prospects for renewable electricity probably lie in South Africa's abundant solar resources, estimated to be amongst the best in the world (Pegels 2010). The national nominal potential for concentrated solar power using existing, proven technology has been estimated as 548 GW, with an effective capacity of 212 GW, which is about five times the country's total electricity generation capacity in 2012—although the technology would have to overcome water scarcity (Fluri 2009). However, recent research has suggested that the energy return on investment for solar photovoltaic power could be in the range of 6–12:1, about half that for coal-fired electricity (Raugei and Fthenakis 2012), and even lower at 2.4:1 in Spain (Prieto and Hall 2013). Wind energy resources are less abundant than solar in South Africa, but the average EROI for wind power in international studies is about 18:1 (Lambert et al. 2012). The South African Wind Energy Association (SAWEA) has estimated that 30 GW of wind power capacity could be installed by private operators by 2025, displacing some 6 GW of conventional base load power, assuming a conservative capacity factor of 0.2 (SAWEA 2010). Although ocean power technology is still in its infancy, wave power potential on the extensive coastline of South Africa has been estimated at between 8,000 and 10,000 MW (Holm et al. 2008).

Although the current financial costs of renewable technologies are mostly fairly high compared to existing conventional power generation, the cost trend for coal- and oil-fired electricity is upwards (as the resources deplete), while the cost trend for renewable energy is largely downward as the technologies improve and increased production yields economies of scale. This is evidenced by the appetite of private sector players, who have initiated solar and wind power projects totalling over 3 GW after the first two successful rounds of bidding under the DoE's Renewable Energy Independent Power Producer Procurement Programme took place in 2012. However, another important factor that needs to be considered in both net energy and economic calculations is that increasing the reliance on intermittent energy sources like solar and wind will require investment in back-up generation and/or storage capacity as well as extensions and upgrades of the national grid. Much more detailed investigation is required before we can adequately evaluate the potential of renewable energy vis-à-vis fossil fuels.

2.3.1.6   Liquid Fuel Scenarios

It is instructive to create scenarios for future liquid fuel supplies in South Africa, based on assumptions and evidence discussed above. As shown in the introduction, several analysts expect world oil production to begin declining within a few years, possibly at a rate of 2–5 % per annum (Hirsch 2008). Furthermore, world oil exports have been stagnant since 2005, and an increasing proportion of these are being consumed each year by China and India (Brown 2013). Thus a conservative assumption is that world oil exports could decline by about 5 % p.a. once global oil production begins to decline, which is assumed here to be in 2015 for illustrative purposes. The simplest assumption for South Africa's oil imports is that they will decline at a similar rate as world oil exports, which assumes that South Africa maintains its share of

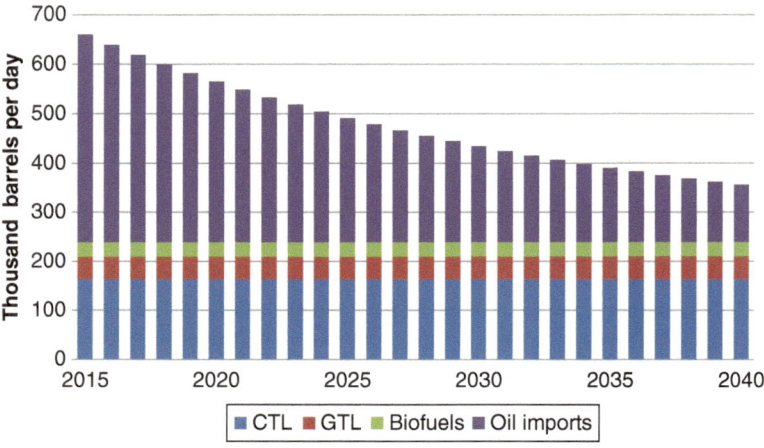

Fig. 2.10 Baseline liquid fuel scenario with 5 % oil import decline rate. *Source*: Author's calculations

world oil consumption, which was about 1 % in 2011. If anything, this is an optimistic assumption, as larger and richer nations are more likely to be able to out-bid South Africans for declining supplies of oil. For domestic synthetic fuel supply, it is assumed for this scenario that Sasol increases its 2012 level of production by 3.2 % in 2014 (to 165 kbpd) as per its stated plans (Sasol 2010) and maintains this level until 2040, assuming that its Secunda facilities commissioned in the early 1980s have an expected lifespan of up to 60 years. Furthermore, it is assumed that PetroSA produces at full capacity (45 kbpd) for 8 years, sustained by the newly found F-O gas field (PetroSA 2010), after which it sources feedstock either from domestic shale gas or from imported LNG. The contribution of biofuels is assumed to be limited to that mandated in the Biofuels Industrial Strategy, namely, 5 % of 2013 road fuels (30 kbdp). Finally, it is assumed that net oil imports rise slightly to 420 kbpd in 2015. Assuming a 5 % decline in oil imports beginning after 2015, the relative contributions of imports and domestic fuel supplies are shown in Fig. 2.10. Total liquid fuel supply would be 74 % of its 2015 level by 2025 and 54 % of its 2015 level by 2040. Should the onset of global oil production decline occur sooner or later than 2015, the depletion profiles would simply be shifted earlier or later by the corresponding number of years.

2.3.1.7   Capital Costs of Alternative Liquid Fuels

Estimated capital costs for various liquid fuel alternatives are listed in Table 2.3. It is clear that CTL and GTL plants are much more expensive than ethanol and biodiesel plants, both in absolute terms and per litre of fuel production capacity. Small-scale biodiesel plants have the lowest cost per litre of daily production. Feedstock costs for all types of plant are highly variable and can be expected to increase over time—as a result of depletion in the case of coal and gas and due to land, water, and food shortages in the case of biofuel crops.

Table 2.3 Comparison of liquid fuel capital costs for some proposed projects

Fuel type	Capital cost	Capacity	Unit capital cost	Source
	R million	litres/day	R/litre/day	
CTL	160,000[a]	12,720,000	12,579	Donnelly (2010)
GTL	74,000[b]	5,247,000	11,206	Engineering News (2011)
Ethanol	2,000[c]	548,000	3,448	Roelf (2012)
Biodiesel	0.085	600	142	Biodiesel Centre (2011)
	0.325	3,000	108	Biodiesel Centre (2011)
	0.025	113	221	NanoElf Biodiesel (2011)
	0.036	200	182	NanoElf Biodiesel (2011)
	164[d]	61,644	2,663	Nolte (2007)

Notes:

[a]Sasol's proposed Mafutha project for the Waterberg area in Limpopo Province, South Africa
[b]Chevron and Sasol's joint-venture GTL plant in Nigeria. Capital costs were converted from US$ to Rands using an exchange rate of R8.8/$ (March 2013)
[c]Proposed ethanol plant to be constructed in the Eastern Cape Province by 2014, as part of the government's biofuels programme
[d]Capital costs cited (Nolte 2007) were for the year 2006; these values were updated to 2011 by using the GDP deflator, which increased them by a factor of 1.37

2.3.2 Energy Conservation and Efficiency

Given the limitations on alternative energy supplies, energy conservation and efficiency are essential to combat the future decline in oil availability and affordability. Since all energy sources are substitutes to some degree—at least in the long run—a national programme to promote conservation and efficiency should apply to all energy sources. The most important area for oil conservation is transport, given that this sector consumes three-quarters of oil products in South Africa; measures to boost efficiency and curtailment of oil use in transport will be discussed in Chap. 3. Consumption of petrochemical products such as plastics can be reduced through improved recycling programmes. The need for bitumen will be reduced to an extent as road-based transport is shifted to railways, alleviating the maintenance pressure on roads (see Chap. 3), although this will come at the cost of coal needed to manufacture steel railways. Efficiency improvements should be implemented at the point of energy generation (e.g. upgrading oil refineries and power plants to be as efficient as possible) as well as distribution (e.g. rationalising the logistics of fuel transport and/or using pipelines rather than trucks to carry fuel). Efficiency measures in the electricity sector could include the adoption of more efficient technologies (e.g. integrated supercritical coal power plants), implementation of smart grids, decentralisation of electricity generation to reduce distribution losses, installation of solar water heaters, and cogeneration of heat and power (Greenpeace 2011; Winkler et al. 2010).

A national drive for energy efficiency and conservation could be led by central government and encouraged by a combination of awareness campaigns, statutory regulations (e.g. mandatory efficiency standards), and economic incentives (e.g. tax rebates for efficiency). Fortunately, several years of electricity supply constraints and rising tariffs have begun to change consumer behaviour, which was wasteful after decades of very cheap power. The "rebound effect"—the tendency of consumers to spend the money saved through energy efficiency gains on other goods and services that embody or use energy—is unlikely to be a major problem given the expected increases in prices for electricity and liquid fuels. The efficiency programme would benefit from explicit national targets for oil demand reduction, similar to the efficiency targets adopted by the Chinese Government (see the further discussion in Chap. 7).

2.3.3 Oil Depletion Protocol

Colin Campbell and Richard Heinberg have suggested an Oil Depletion Protocol as a cooperative response to declining oil supplies (Campbell 2006; Heinberg 2006a). The Protocol in essence requires all oil-importing nations to agree to reduce their annual oil imports by a percentage equal to the World Oil Depletion Rate, which has been estimated by Campbell as approximately 2.6 % per annum. In addition, oil-producing nations would agree to reduce their rate of production by their National Depletion Rate. The result will effectively be a global rationing system, which is intended to help stabilise oil prices and avoid wars over remaining oil and thereby ensure that economic and social conditions are more conducive to the crash programme of mitigation required to avoid the worst potential economic impacts. Heinberg suggests that the Oil Depletion Protocol could operate alongside carbon emission-based agreements such as a strengthened and extended Kyoto Protocol.

While the Oil Depletion Protocol has merits in theory, it would face similar obstacles to international adoption and implementation as have confounded climate treaty negotiations. In particular, there are likely to be conflicts between various groupings of countries, such as between developed and developing nations (the latter may argue that they have a right to a greater proportion of remaining oil reserves to compensate for their lower historical oil consumption) and between oil-importing and oil-exporting nations. The situation resembles a complex version of the prisoner's dilemma, in that the individually rational country strategies are likely to lead to a socially suboptimal outcome. The Protocol represents a mutually cooperative set of strategies that would be very difficult to achieve in practice but would yield benefits for most countries.

2.4 Conclusion

All of the potential substitutes for imported oil have both advantages and limitations. The main advantage of CTL and GTL technologies is that they produce fuels that can be used in existing transport infrastructure. The disadvantages include a depleting resource base, probably relatively low EROI, high capital costs, water scarcity, and pollution (including GHGs). Biogas has potential as a sustainable, local replacement for LPG and wood fuel. Liquid biofuels (ethanol and biodiesel) might make a small contribution to liquid fuel needs but will be severely constrained by scarcity of water and high-quality arable land and may undermine food security. The best prospects for biofuels are probably for small-scale, decentralised biodiesel production, especially for use on farms. Electricity is a flexible energy carrier that can be used for multiple purposes including transport, although new generation, transmission, and distribution infrastructure will need to be constructed. Nuclear energy provides a relatively reliable base-load power source but faces very high capital investment and decommissioning costs, risks of contamination, and the as-yet unsolved issue of long-term waste disposal. Electricity generated from renewable sources has several limitations that need to be overcome, including intermittency, low power density, and relatively high costs for solar power compared to coal-fired electricity (at least when the latter excludes external environmental and social costs). These are merely preliminary observations; we still require detailed analyses to determine the life cycle net energy returns of all of the alternative energy sources, including their supporting capital infrastructure.

Owing to the constraints, costs, lead times, and risks attached to developing alternative energy sources, a nationwide programme of energy conservation and efficiency is imperative to address the oil supply challenge. In fact, conservation should be the first priority, since it offers opportunities to capture "low hanging fruit" that are cheaper and easier to implement and can create time and budget space for constructing new infrastructure to deliver alternative energy sources in the longer term. The next chapter explores the potential for fuel savings in the transport sector, which consumes the bulk of oil products in South Africa.

Chapter 3
Transport

Effective transport systems are essential for the conduct of local, regional, and international commerce and trade, and mobility is an important determinant of human welfare in the modern world. South Africa has a transport infrastructure that is the envy of many developing countries (ASPO-SA et al. 2008b). An extensive road network, comprising both paved and (to a much greater extent) unpaved roads, covers much of the country. A rail network of 20,000 km connects all of the major urban settlements in the country, and approximately 80,000 wagons and 2,300 locomotives operate on this network. The South African Rail Commuter Corporation (SARCC) operates commuter railways in the six major metropolitan areas, but there are no light railways or underground rail systems in South Africa's cities. South Africa has in excess of 20 commercial airports, although many of these are small and provide limited services (ASPO-SA et al. 2008b: 24). The majority of international flights land at the O.R. Tambo International Airport (ORTIA) in Johannesburg, while airports in Cape Town and Durban also offer some international flights. South Africa has a coastline of 2,954 km, on which there are 18 notable ports including eight multipurpose commercial ports. These ports are connected to the rail and road networks and serve as entry and exit points for internationally traded goods.

Yet despite—and in some cases because of—this infrastructure, South Africa's transport system suffers from numerous deficiencies and vulnerabilities, especially in its uneven provision of services across the population and the extremely high level of dependence on oil. This chapter begins by detailing this oil dependence and identifying the key strengths and vulnerabilities of the transport system. It then discusses the likely impacts of oil price and supply shocks on the mobility of people and goods, before considering a wide range of mitigation options for conserving fuel and switching to less oil-dependent modes of transport.

J.J. Wakeford, *Preparing for Peak Oil in South Africa: An Integrated Case Study*,   35
SpringerBriefs in Energy, DOI 10.1007/978-1-4614-9518-5_3, © Jeremy J. Wakeford 2013

3.1 Oil Dependence of the Transport System

Despite the country's extensive transport infrastructure, nearly one-half of South Africa's population still rely on non-motorised transport (NMT) for their mobility (DoT 2005). The other half of the citizenry make use primarily of road-based motorised transport, including private motor vehicles and minibus taxis (MBTs). The reliance on motorised transport is substantially higher in metropolitan and urban areas compared to rural areas for almost all transport modes. According to the most recent survey data, the National Household Travel Survey, South Africa had some 10 million commuters in 2003 (DoT 2005). At this time the country's commuters were overwhelmingly dependent on road transport, including MBTs (22 % of commuters), cars (15 %), buses (6 %), and other taxis (3.7 %), while just 2.3 % of commuters reportedly used trains. Since that survey was conducted, car ownership and the number of registered MBTs have increased significantly. As of 31 December 2012, there were approximately 9.5 million self-propelled registered vehicles on South Africa's roads, including 6.1 million motor cars (representing approximately 118 cars per 1,000 people in a population of 51.8 million), 356,000 motor cycles, 286,000 MBTs, and 52,000 buses (eNatis 2013). Meanwhile, the use of passenger trains has been relatively static as rolling stock capacity was not expanded. Passengers in general have a low regard for public bus and rail transport systems as a result of perceptions of long distances between dwellings, stations, and bus stops, and "[t]he public transport system is underutilised, severely underdeveloped and undercapitalised in relation to commuter needs" (ASPO-SA et al. 2008b: 9). In recent years, however, the development of bus rapid transit (BRT) systems has begun in Johannesburg and Cape Town, with similar systems planned for at least three other major cities. Water-based passenger transport is negligible, as the country has no significant navigable rivers or canals and no passenger service along the coast apart from tourist cruise ships.

Freight transport is also heavily road based, relying on a stock of over 2.1 million light commercial vehicles and 342,000 heavy trucks as of December 2012 (eNatis 2013). In 2012, road freight accounted for 88.5 % of freight tonnage and rail for the remaining 11.5 % (see Fig. 3.1). Mining products, such as coal and iron ore, account for almost half of all freight tonnage, manufacturing for another 45 %, and agricultural products for just 6 % (GCIS 2007: 578). Since the mid-1990s, rail freight volumes on corridors have declined while road freight volumes have grown steadily, apart from a dip in the recession year of 2009. Logistics costs in 2012 were estimated at 12.8 % of GDP, approximately half of which were transport costs (CSIR 2013). Relatively small volumes of high-value freight are transported by air. South Africa makes very limited use of waterborne freight transport inland but relies heavily on international shipping for exports and imports.

An examination of energy consumption by transport gives an even clearer indication of the sector's dependence on oil. The transport sector utilised 83 % of all petroleum energy, 2 % of all electricity, and 25 % of total final energy consumed in South Africa in 2009 (IEA 2013). Within the transport sector itself, 98 % of the

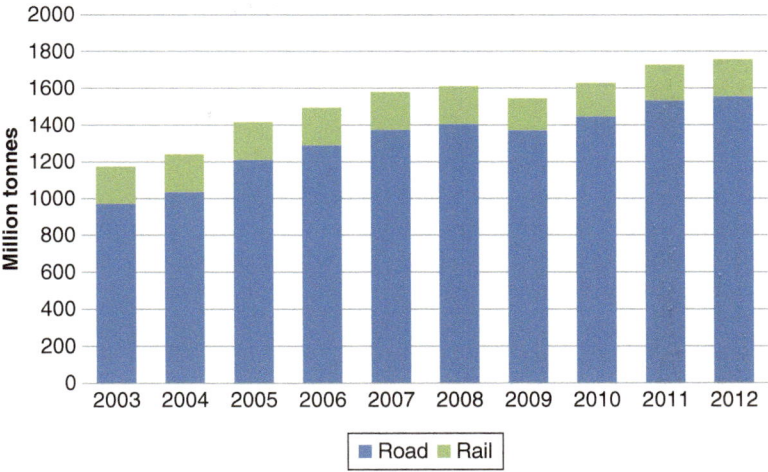

Fig. 3.1 Land freight transport in South Africa, 2003–2012. *Source*: CSIR (2013)

Table 3.1 Energy consumption by the transport sector, 2009

Transport sector	Coal		Petroleum		Electricity		Total	
	TJ	%	TJ	%	TJ	%	TJ	%
International aviation	–	–	36,118	4.0	–	–	36,118	3.9
Domestic aviation	–	–	47,559	5.3	230	1.7	47,789	5.2
Road	–	–	711,364	79.1	86	0.7	711,450	77.8
Rail	2,035	100.0	1,788	0.2	10,442	79.3	14,265	1.6
Pipelines	–	–	–	–	355	2.7	355	0.0
Internal navigation	–	–	102,857	11.4	195	1.5	103,052	11.3
Non-specified	–	–	–	–	1,856	14.1	1,856	0.2
Total	2,035	100	899,684	100	13,165	100	914,885	100

Source: DoE (2012)

energy consumed is derived from petroleum products, and only 2 % from electricity, and these shares have hardly changed since 1990. Clearly, the transport sector is overwhelmingly dependent on liquid petroleum fuels, i.e. petrol, diesel, and jet fuel. Table 3.1 shows the consumption of energy disaggregated by energy carrier and by transport mode. Road transport dominates transport energy consumption with 79 % of petroleum fuels and 78 % of all energy. Aviation (including international and domestic air transport) accounted for 9.3 % of petroleum consumption within the transport sector in 2009, while 11.4 % was reportedly consumed for internal navigation (this figure is of questionable reliability given the lack of navigable rivers). Rail used predominantly electricity (79 % of the power consumed by the transport sector) plus a small amount of diesel and coal but less than 2 % of total energy consumed by transport.

3.2 Vulnerabilities and Likely Impact of Oil Shocks on the Transport System

Transport is clearly the sector with the greatest vulnerability to diminishing oil supplies. The major vulnerabilities to oil shocks inherent in South Africa's transport system include infrastructural, modal (passenger and freight), and geographical dimensions, which are explored in turn below.

Transport infrastructure in South Africa, although extensive relative to many other developing countries, suffers from several problems. A major weakness of the road network is the deficit in maintenance in many areas, with an estimated backlog of R75 billion (approximately $10 billion) in April 2010 (Davenport 2010). Some four-fifths of the country's roads were older than the 20-year lifespan for which they were designed, while "[t]he percentage of bad and very bad roads in the secondary road network of South Africa increased from 8 % in 1998 to 20 % in 2008" (CSIR 2010: 5). Road maintenance costs are vulnerable to oil price rises, since the bitumen used for surfacing paved roads is derived from crude oil, as is the diesel fuel that is used to power road-building vehicles and machinery. According to the South African National Roads Agency Limited (SANRAL 2010: 15), a "spiralling cost of bitumen" along with rising fuel prices was the major reason underpinning the upward trend in road construction prices in the 2000s, considerably above the rate of consumer price inflation. Diminishing affordability of road maintenance would lead to an increased rate of deterioration of road surfaces and "drastic increases in vehicle maintenance and repair costs" for companies and private vehicle owners as well as an increase in road accidents (CSIR 2010: 24). Similar concerns apply to the maintenance of airport runways, although with even more critical safety implications.

Although the rail network has significant spare capacity for moving freight, it too has been neglected in terms of maintenance and upgrading (ASPO-SA et al. 2008b: 50–51). Much of the nation's rolling stock is very old and needs to be replaced. The well-developed port infrastructure could serve the economy well if more freight is shifted from road to sea. However, since manufacturing production is concentrated in Gauteng, there is limited scope for such a modal transfer, especially considering the lack of intermodal facilities. Many of the country's airports were upgraded or expanded in preparation for the FIFA Soccer World Cup in 2010, but this expenditure will be of doubtful use for the future when air travel is likely to be severely constrained internationally and domestically by rising fuel costs. Public transport in several major cities also received a significant boost as part of the preparations for the World Cup, when construction began on BRT systems in Johannesburg and Cape Town. In general, the adverse economic impacts of peak oil are likely to constrain capital spending on transport infrastructure just when they are needed most.

The mobility of the more than half of South Africa's population that relies on motorised transport is highly vulnerable to oil price shocks, given the overwhelming reliance on liquid petroleum-fuelled vehicles. Poorer transport users tend to spend a much higher proportion of their incomes on transport and therefore are more vulnerable to rising fuel and transport costs than their wealthier counterparts. Public transport users will face increasing fares, and if these become unaffordable, commuters may have to walk or even be unable to travel to their places of work. Motorists will respond to higher fuel prices with a range of behavioural adaptations, starting with reduced discretionary driving and later shifting to public transport, which may quickly become oversubscribed. While air travel will be particularly vulnerable to rising fuel prices (since fuel accounts for a relatively high proportion of total costs), business travellers can adapt to some extent by telecommuting. Many of those travelling by air for other purposes (e.g. tourism) are likely to have to shift mode and/or reduce their travel distances and/or frequency. Airlines will take increasing financial strain and may have to terminate services on low-volume, short-haul routes. In fact, several small airlines operating domestic routes have already gone bankrupt in the past few years, and the state-owned national carrier South African Airways has been bailed out repeatedly with billions of rands of taxpayer money.

The heavy reliance of freight transport on trucks presents a major challenge to the economy, both in terms of future fuel supply constraints and the impact of rising costs. A wide range of vulnerabilities characterise freight transport in South Africa, including inadequate technology, equipment, and facilities; outdated infrastructure; lack of intermodal facilities; capacity bottlenecks; monopoly ownership of certain infrastructure; skill shortages; and a lack of information (Lane 2009). Given the overwhelming dependence of freight movement on road transport, rising fuel prices would have a significant impact on the trend in freight and logistics costs. It has been estimated that a tripling in the fuel price would raise logistics costs by 53 % under conditions pertaining in 2008 (CSIR 2010: 16). As stated in the annual State of Logistics Survey of 2012, "[t]he vulnerability of transport costs to a volatile exogenous cost driver—the price of crude oil—and South Africa's entrenched dependence on road transport does not bode well for the economy if the future is to be business-as-usual" (CSIR 2013: iii). Some industries and retailers might absorb a portion of the added freight costs internally, but in general higher freight costs will be passed along the value chain to final consumers in the form of higher retail prices for goods. Businesses with a high degree of reliance on freight movement (i.e. where transport costs are a significant proportion of the final costs of goods) will find that rising fuel costs steadily erode their profit margins. For international trade, rising fuel costs will act like a tariff barrier, favouring local trade over long-distance trade (Rubin 2009). Being highly energy intensive, air freight is much more sensitive to fuel prices than other modes of freight transport. In the long term, costs of air freight may become prohibitively high, resulting in the collapse of markets for all but the highest value goods.

Transport vulnerabilities can also be identified on a geographical basis. For one thing, there are substantial distances between major metropolitan areas and towns; the two largest cities—Cape Town and Johannesburg—are 1,400 km apart. The bulk of liquid fuels are consumed in metropolitan areas due to the high concentration of vehicles found there and the phenomenon of urban sprawl. Areas in the hinterland that have no rail access are also highly vulnerable, with a danger that both people and assets could become stranded as many parts of the country are accessible only by roads (ASPO-SA et al. 2008b: 56). Both industry and population are concentrated in the interior of the country, far from ports (and at a much higher altitude), which means that two long-distance corridors (Durban–Gauteng and Cape Town–Gauteng) carry large freight volumes. Finally, South Africa is very far from most of its trading partner countries, which makes international trade and tourism especially vulnerable to rising transport costs.

Short-term, sudden shortages of fuel could have a very large impact on transport. For example, a sudden interruption of jet fuel supplies could have serious implications as the Department of Minerals and Energy warned that "indications are that there is not enough space in South African airports to park all airplanes that are, at any one time, heading for or in South Africa" (DME 2007a: 21). For road passengers (whether using private or public vehicles), localised fuel shortages could result in disabling immobility. Shortages are likely to result in extensive queuing and possibly even conflict at filling stations, while hoarding responses are also possible. In the case of freight, fuel shortages would result in disruptions to logistics chains. The longer the production chains involved, the greater the potential for disabling disruptions. Just-in-time delivery systems are particularly at risk of logistics failures, with the result that shortages of various commodities could arise at the retail level. If fuel shortages are severe, localised shortages of food could arise within a few days.

3.3 Mitigating Peak Oil in the Transport Sector

Mitigating peak oil in the transport sector requires a reduction of the overwhelming dependence on petroleum fuels of the various modes of passenger and freight transport, especially motorised road and air transport. The national Department of Transport is aware of the peak oil issue; its *National Transport Master Plan 2005–2050* incorporated a set of reports on the implications of global oil depletion for transport in South Africa (ASPO-SA et al. 2008a, b, c, d). Encouragingly, the DoT has recently given renewed attention to railways, including investigations into the feasibility of constructing high-speed rail lines on major transport corridors and upgrading passenger rail infrastructure. The Gautrain rapid rail service, which connects the metropolitan centres of Pretoria and Johannesburg, and links with the OR Tambo International Airport, was completed in 2012. As previously mentioned, integrated (bus) rapid transit systems are being rolled out in several major cities. Nonetheless, very large amounts of money have been spent in recent years on oil-related transport infrastructure such as roads and airports. Fortunately, there is much that can be done to reduce the transport sector's reliance on oil, including demand-side management, a switch to more fuel-efficient vehicles, and modal shifts.

3.3.1 Demand-Side Management

There is a wide array of measures than can reduce fuel consumption significantly in the short to medium term and which do not require substantial financial outlays for

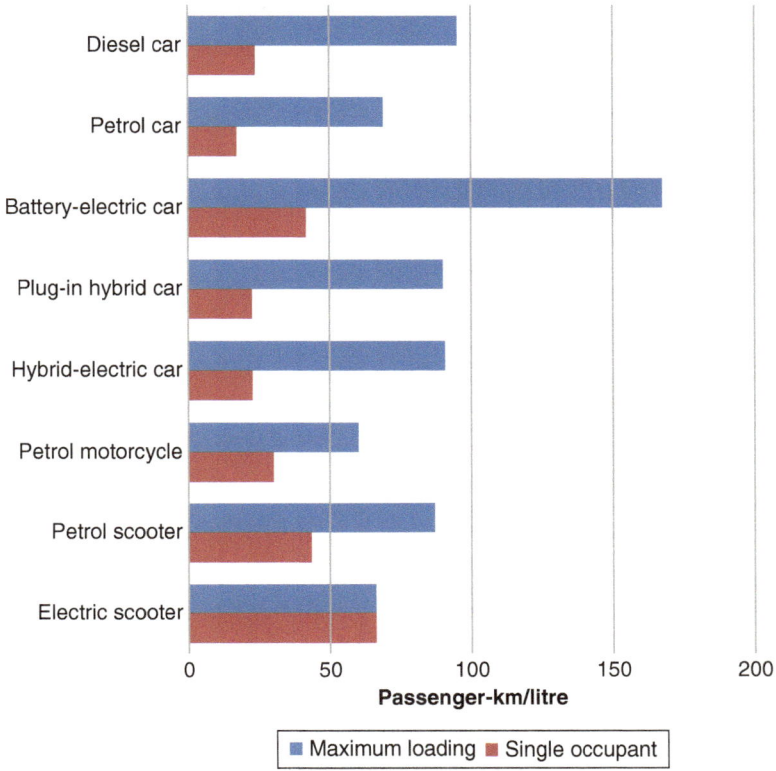

Fig. 3.2 Energy efficiency of passenger road vehicles. *Source*: Wakeford (2012)

infrastructure. The simplest way to save fuel is through eco-driving techniques, such as use of correct gears, avoiding unnecessary acceleration and braking, appropriately inflating tyres, curbing the use of air conditioning, driving with windows closed, and adequately maintaining vehicles. These measures, which can save up to 5 % of fuel (IEA 2005), can be encouraged by public information campaigns and driver training. Although individual behaviour is not easy to change, rising fuel prices will provide a strong economic incentive for the adoption of fuel economy measures. Second, traffic management systems can reduce fuel consumption by between 5 and 15 % by helping to ensure that vehicles travel at more efficient speeds (Vanderschuren et al. 2008: 25). One of the most effective ways to reduce fuel consumption in road vehicles is to reduce maximum road speed limits (IEA 2005), for example, to 90 km per hour on highways. Success, however, will depend on adequate enforcement (which is difficult enough with current—higher—speed limits) and will require expenditure on new signage and possibly extra law enforcement personnel. Third, the greatest opportunity for achieving energy efficiencies in passenger transport lie in boosting vehicle occupancy rates (see Fig. 3.2). This applies mainly to private cars but also to buses, trains, and airplanes. Car-pooling (or ride-sharing) aims to reduce the prevalence of single-occupant private vehicles by encouraging drivers to take passengers. Authorities can promote car-pooling by establishing car-pool or high-occupancy vehicle lanes on urban freeways, designating park-and-ride lots, introducing Internet-based systems to match riders, levying congestion charges in city centres, and conducting awareness campaigns (IEA 2005). However, widespread car-pooling could face obstacles in a South African context, for example because of high crime rates and a threat of hijacking. A fourth demand management option is for local governments to partner with businesses to encourage telecommuting and flexible work schedules (including compressed work weeks and staggered starting and ending times), which reduce the need for commuting and reduce traffic congestion. In the case of freight transport, the most cost-effective measures for reducing oil consumption are those requiring little new infrastructure and which can be implemented relatively easily in the short to medium term, such as improved vehicle maintenance, optimised routing and scheduling, and intelligent traffic management solutions (Lane 2009).

3.3.2 More Efficient and Alternative-Fuel Vehicles

Internal combustion engine vehicle (ICEV) technology has not changed fundamentally in over a century and is highly energy inefficient, using approximately 17 % of the energy contained in the fuel to propel the vehicle and less than 1 % to move a single occupant (Lovins et al. 2005: 46). An effective way of reducing fuel consumption over the medium to longer term is therefore to incentivise consumers and businesses to buy more efficient motor vehicles—e.g. smaller, lighter models. This can be achieved through the introduction of a "feebate" system, whereby extra taxes are imposed on larger, gas guzzling vehicles while rebates are given on purchases of more fuel-efficient models. The National Treasury has already introduced a carbon emission tax on new motor vehicles, which encourages consumers to buy more efficient cars. The government could also impose fuel efficiency standards on vehicle manufacturers, as has been done in the United States and Europe, but only about 30 % of domestic car sales are locally produced. Improvements in vehicle design (e.g. enhanced aerodynamics and friction management) can also boost fuel efficiency in other modes of transport, such as air and rail.

Even greater reductions in liquid fuel use can be achieved through the replacement of ICEVs with vehicles powered by alternative fuels and propulsion mechanisms, many of which are already commercially available. For the most part, these new vehicles are for private passenger transport, although increasingly some of the technologies are being used in buses and even trucks. These technologies include liquefied petroleum gas (LPG) vehicles, compressed natural gas (CNG) vehicles, compressed air vehicles (CAVs), hydrogen-powered vehicles (HPVs), hybrid electric vehicles (HEVs), plug-in hybrid vehicles (PHVs), and battery electric vehicles (BEVs). Some have better prospects than others. LPG vehicles offer no respite from oil dependency and only modest efficiency gains. CNG, which is used extensively

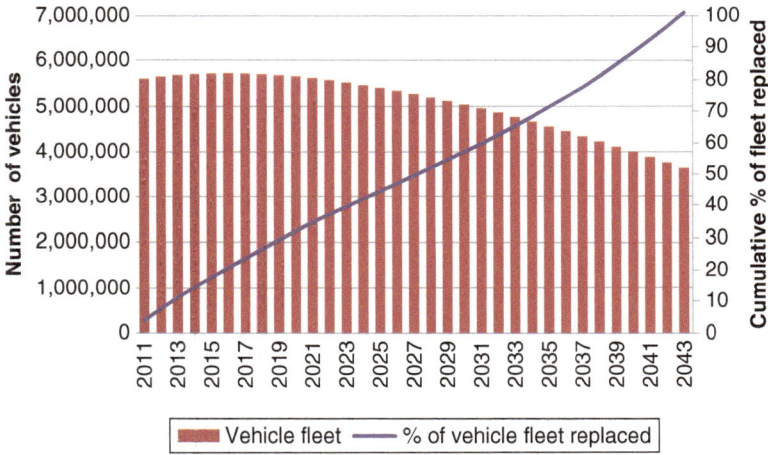

Fig. 3.3 Cumulative replacement of passenger vehicle fleet assuming a 5 % annual decline in new vehicle sales. *Source*: Author's calculations

for transport in developing countries like Pakistan, India, and Iran, would only be feasible in South Africa if competitively priced gas from shale basins or neighbouring countries were to become available. However, it would be more efficient to convert the gas into electricity and use this to run electrified mass transit (Kendall 2008: 132). CAVs and HPVs have yet to be successfully commercialised, possibly because of inherent energy efficiency losses (Creutzig et al. 2009; Strahan 2007). By contrast, HEVs, PHVs, and BEVs are now being manufactured by a slew of major automobile companies. Hybrids can be at least twice as fuel efficient as comparable ICE cars but are still dependent on oil and the price tag is relatively high. BEVs are independent of petroleum fuels and are highly energy efficient (see Fig. 3.2) but have limited range and the batteries currently rely on scarce lithium reserves. Major advances in battery technology are likely required for widespread deployment of BEVs on anything like the current scale of automobile usage. PHVs offer similar energy efficiency to HEVs and longer driving ranges than BEVs. The major drawback of PHVs is their greater cost, although this might fall as manufacturing economies of scale are exploited.

There are two major constraints facing a large-scale replacement of ICEVs with alternative vehicle technologies: (1) it will take several years for vehicle manufacturers to ramp up production capacity from today's negligible level; and (2) consumers face affordability constraints. Vehicles represent a very large expenditure item for households and businesses, which in a post-peak future could face increasingly adverse economic circumstances and falling discretionary incomes and profits, respectively. Figure 3.3 projects the total passenger vehicle stock and the percentage replacement by new vehicles, assuming a 5 % annual decline in vehicle sales. The decline in sales is based on the assumption that GDP and household income will contract, and/or fuel and vehicle prices will rise, as global oil supplies decline. Assuming that 3 % of the 2010 vehicle stock is scrapped every year (scrappage rates are low due to the extensive second-hand market), it would take 33 years to replace the entire passenger vehicle fleet with new, more efficient, vehicles. This replacement would also require a large capital expenditure on the part of households and firms—money that could perhaps be spent more effectively if directed to efficient mass transit.

3.3.3   Modal Shifts

The quantitatively largest opportunities for reducing oil dependence are presented by modal shifts. For passengers, the primary modal shifts are from private road vehicles to NMT, electric bicycles, scooters and motorcycles, and collective public transport such as minibuses, buses, and various grid-connected vehicles (GCVs) such as trams, trolley buses, and trains. In addition, shifting from air travel to other modes (e.g. rail) is applicable to a numerically small but financially significant proportion of travellers. Unfortunately, most of these shifts would likely be regarded by citizens as a decline in affluence. For freight, the main modal shifts are from road and air to rail (which should be progressively electrified). In general, government can employ two strategies to encourage modal shifts: (1) use fiscal measures (e.g. congestion charges, vehicle taxes, or fuel taxes) that make it more expensive for people to use road vehicles and (2) provide alternative modes of transport that are sufficiently attractive, e.g. invest in public transport and freight rail infrastructure.

In urban areas, oil dependence can be reduced through a modal shift from motor vehicles to NMT—namely, walking and cycling, which are the most energy-efficient and least costly forms of passenger transport. Other benefits of NMT include reduced air pollution, health benefits associated with physical exercise, enhancement of community cohesion, and less use of land and road space (DoT 2008b: 10). A shift to NMT can be encouraged through the provision of safe pedestrian walkways and cycle lanes (which should link to public transport facilities where possible), bicycle hire schemes, and fiscal measures that discourage car use. Nevertheless, NMT faces both practical and cultural constraints in South Africa, where cities are characterised by urban sprawl and the location of informal or low-income settlements far from urban centres. The Department of Transport recognises that public resistance to a shift from private vehicles to NMT may present an obstacle since vehicles are regarded as enabling independence and as a symbol of status. Another major challenge will be ensuring the safety and security of pedestrians and cyclists, since South Africa has notoriously high rates of traffic fatalities and crime. In poor rural and urban areas, the majority of the population already relies on NMT, especially walking. Here the opportunity is to improve mobility, for example by subsidising bicycles or expanding public transport facilities such as buses.

Another modal shift option is for commuters to make more use of electric bicycles and scooters, which offer larger travel ranges than bicycles but are still highly energy efficient. Since affordability of EBs would be constrained for low-income households, there is a case for government subsidies funded by higher road vehicle

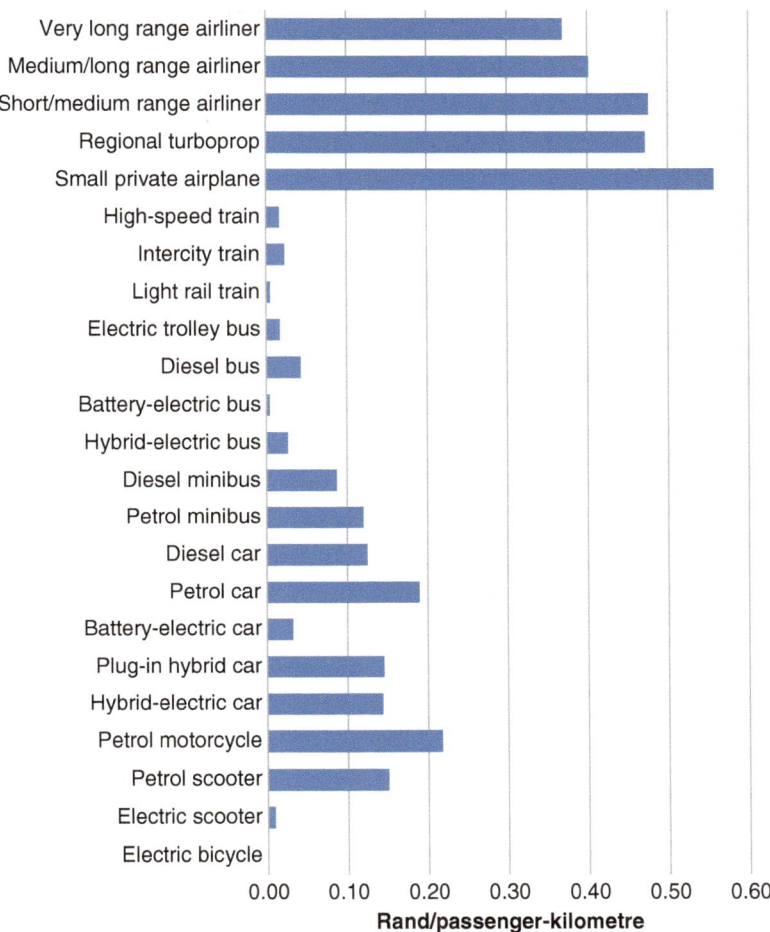

Fig. 3.4 Energy cost of various passenger transport modes under maximum loading. *Source*: Updated from Wakeford (2012). *Notes*: (**a**) Energy costs: Petrol = R 13.08/l (Gauteng Unleaded 95, 6 March 2013); diesel = R11.92/l (Gauteng 0.005 Sulphur, 6 March 2013); electricity = R 0.61/kWh (Eskom average national tariff in March 2013); jet fuel = diesel cost (assumed). (**b**) Maximum loading is assumed to be four for passenger cars and two for motorcycles and scooters

taxes. Even petrol-driven scooters and motorcycles have several advantages over cars, including higher fuel efficiency (compared to single-occupant cars), lower capital costs, lower maintenance costs, and less usage of land and road space. Currently, motorcycles comprise just 3.7 % of motorised road vehicles, indicating much potential for growth in market share. By contrast, motorcycles are more common than cars in many South and East Asian countries. The main disadvantage of scooters and motorcycles in a South African context relates to the poor level of safety on roads.

MBTs provide the bulk of public transport in South Africa. A fully loaded diesel minibus is a comparatively efficient mode of road transport in terms of cost per passenger kilometre (see Fig. 3.4). Such vehicles are flexible in terms of route and especially schedule, which allows them to maximise the number of passengers they carry and hence their energy efficiency. The major constraint on shifting passengers from private vehicles to MBTs is arguably perceptions of safety, since the MBT industry in South Africa has a notorious history of violence and road fatalities. Normal bus services have operated in major cities for many years, but occupancy rates—and therefore energy efficiencies—are highly variable. Recognising the potential fuel savings offered by efficient, well-utilised bus services, several metro-politan authorities have recently begun rolling out BRT systems. Ideally the vehi-cles should be hybrid electric or fully electric models so as to reduce dependency on petroleum fuels. The infrastructure costs of BRT systems are an order of magnitude cheaper than those of light rail (e.g. R14 million per kilometre compared to R140 million/km, respectively), can be built much faster, and are more flexible as buses need not stick to trunk routes (Hook 2009).

For similar load factors, trains are more energy efficient than buses, cars, and aeroplanes (Schiller et al. 2010). After decades of neglecting the passenger rail ser-vice, the Department of Transport announced a comprehensive rail investment pro-gramme in 2010 as part of the National Transport Master Plan, aimed at upgrading existing stock, acquiring new rolling stock, and developing new rail corridors (Mahlalela 2010). R123 billion (about $12 billion) has been budgeted by the gov-ernment over the coming 18 years for the upgrade of passenger rail rolling stock; but this amount is trivial compared to the roughly R65 billion spent by households on private passenger vehicles *each year* (SARB 2013). Taxes on vehicle sales and subsidies for public transport can address this gross imbalance and inefficiency in societal expenditure patterns. However, successfully attracting passengers from cars to public transport will also require improvements in the provision of public trans-port services, in terms of speed, reliability, regularity, safety and security, conve-nience, comfort, and costs. Intercity passenger rail services are very meagre at present but need to be expanded to accommodate a shift from air travel, which is likely to become progressively less affordable in a post-peak oil future.

GCVs are powered by electricity that "is generated remotely and delivered directly by wire or rail to the motor as the vehicle moves" (Gilbert and Perl 2010: 3); examples include heavy rail, light rail, trams, and trolley buses. GCVs have some major advantages: they have a proven track record spanning over a century, being used for intra- and intercity travel in many countries; they are the most energy efficient form of motorised transport; and they do not require heavy and costly bat-teries as do BEVs. For example, trolley buses in the United States consume on aver-age 0.85 megajoules per person-kilometre (MJ/pkm), while diesel buses use 2.40 MJ/pkm (Gilbert and Perl 2008: 156). Disadvantages include large infrastruc-ture requirements (constructing the grid) and hence high initial capital costs and the fact that mobility is limited to the grid's coverage (although in some instances bat-teries could be added to allow limited off-grid travel). Nevertheless, some experts see GVCs as forming the backbone of future land-based transport systems (Gilbert and Perl 2008; Kendall 2008).

In the case of freight, there is great scope for shifting bulk loads from trucks to railways along the main corridors such as Johannesburg–Cape Town and Johannesburg–Durban. Transnet's R300 billion capital expenditure programme is welcome, and although much of this is geared towards expanding mineral export lines, the new locomotives could be utilised at a later stage for general freight. Fortunately, Transnet is purchasing locomotives that can operate on diesel fuel or electricity, providing the greatest flexibility and possibilities for energy efficiency and oil independence.

3.3.4 Energy Costs of Transport Modes

Figure 3.4 provides a comparison of energy costs for a wide range of transport modes and vehicles, based on (mostly) international figures for transport energy efficiency and local energy prices. Electric bicycles have the lowest energy cost per passenger kilometre (p-km), followed by light rail trains and battery electric buses. Other types of rail, as well as diesel and hybrid electric buses, also have very low energy costs when fully loaded. The energy costs of air transport are much higher than for all other transport modes. It is important to note that the relative energy cost discrepancies will be exaggerated should liquid fuel prices rise faster than electricity prices. Not included in these figures, but nonetheless very important, are the embodied energy costs of the various transport vehicles and their supporting infrastructure (e.g. roads or railways)—clearly a balance has to be struck between capital cost and running costs.

3.4 Conclusion

South Africa's transport system is overwhelmingly dependent on oil and is dominated by relatively inefficient road vehicles for moving both passengers and freight. The government and its transport parastatals are taking some steps in a more sustainable direction by investing in BRT systems and upgrading passenger and freight rail services. However there is still too much expenditure on roads and airports that continue to lock the country into an oil-dependent pathway. Fortunately, the transport sector offers many avenues for reducing oil consumption. In the short term, a wide range of demand management measures aimed at improving energy efficiency can be implemented cost effectively both to cope with immediate supply or price shocks and to create the space and time needed for longer term infrastructure investments. Walking and cycling could be promoted as more sustainable and cheaper alternatives in dense urban areas, while bicycles offer improved mobility for rural residents. As far as motorised mobility is concerned, Gilbert and Perl (2010: 2, 4) argue that "[o]ver the next two or three decades motorised land transport will become mostly propelled by electric motors" and describe electricity as "the ideal transport fuel for an uncertain future". This is because (1) electricity can be derived from a wide range of primary energy sources, including renewables; (2) transport systems based on electricity can easily adapt to changing primary energy sources and thus avoid the need for changing infrastructure that is dependent on a particular energy source (e.g. oil or natural gas); and (3) the transport system's energy requirements will not constrain innovation in energy production systems. Electrification of transport will involve the replacement of ICEVs with BEVs and GVCs, including modal shifts from road and air to rail. Remaining diesel and biodiesel should be prioritised for heavy uses with less flexible fuel options than personal transport, such as freight trucks and buses (City of Portland 2007: 39) as well as air transport. Transport electrification will be time-consuming and costly (requiring enhanced/smarter electricity grids) but also provides a transition pathway for energy from oil to renewable electricity. Nevertheless, more research is needed to provide a detailed evaluation of potential costs and benefits of both public transport and freight rail (e.g. in terms of life cycle energy usage and economic costs of rolling stock plus its supporting infrastructure, measured per passenger-kilometre or tonne-kilometre, respectively).

The next chapter considers the sector of the economy with the second highest level of oil dependency, namely, agriculture.

Chapter 4
Agriculture

Agriculture is a quantitatively small but qualitatively vital sector of the economy. It represents the base layer of the economy in that the population and labour force must be fed in order to engage in other economic activities. Since economic and social stability depend on a healthy, functioning system of agricultural production and food distribution, the threats posed to these by peak oil need to be taken very seriously.

South Africa has a total land area of 127 million hectares, of which just over 100 million hectares (82 %) is classified as farmland (DAFF 2012). The vast majority of this farmland (84 million hectares or 69 % of the total land area) is suitable for grazing only. Because South Africa resides more or less at the latitudes of descending air masses of the Hadley cell circulation, only 16.7 million hectares (14 %) of the country's land area receives sufficient rainfall to be potentially arable, and only about a fifth of this land is of high quality (GCIS 2009: 47). Water scarcity is a limiting factor for agriculture, and just 1.35 million hectares (1.5 % of agricultural land and less than 10 % of arable land) is under irrigation (DAFF 2012).

South Africa's agricultural economy is made up of two parts: an industrialised commercial sector and a largely rural subsistence or smallholder sector (GCIS 2009: 47). Commercial farmers account for at least 95 % of total marketed agricultural produce (Food and Agriculture Organisation (FAO) 2005: 2). The commercial agriculture sector produces a wide range of commodities, including livestock products (meat and dairy products), field crops (grains such as maize, wheat, and sorghum; sugar; oil seeds; and cotton), and horticultural produce (fruits and vegetables). Maize occupies half of all the land under crops (FAO 2005: 13), is the most important food crop by volume of output, and is the staple food for the majority of South Africans. Subsistence farming occurs predominantly in the rural, former "homeland" areas of South Africa (Pauw 2007: 196) and contributes less than 5 % of total agricultural output (FAO 2005: 2). Subsistence farming involves a small share of the South African population relative to other sub-Saharan African countries, where it remains a major contributor to livelihoods (Baiphethi and Jacobs 2009). Nevertheless, there are approximately four million South Africans involved in subsistence farming, mostly to secure an "extra source of food" (Aliber and Hart 2009: 439).

J.J. Wakeford, *Preparing for Peak Oil in South Africa: An Integrated Case Study*, SpringerBriefs in Energy, DOI 10.1007/978-1-4614-9518-5_4, © Jeremy J. Wakeford 2013

The first section of this chapter describes the extent of the South African agriculture sector's dependence on oil and discusses the issue of national food security. The second section summarises the anticipated impacts of oil shocks. The third section explores strategies to mitigate peak oil by reducing reliance on oil inputs and relocalising food production and consumption, while the final section provides a brief summary.

4.1   Oil Dependence of Agriculture

The agriculture, forestry, and fishing sector accounted for 3 % of total final energy consumption in 2008, which was similar to its 2.9 % contribution to gross domestic product (IEA 2013; SARB 2013). As seen in Fig. 4.1, the relative contributions of coal, petroleum, and electricity to total energy consumption in agriculture have not changed substantially over the past two decades, although the already small share of coal has diminished further. In 2008, more than two-thirds (69 %) of the energy used by the agricultural sector was in the form of liquid petroleum fuels, while electricity contributed 30 % and coal just 1 % directly (although 90 % of electricity is coal fired). Energy and oil intensity varies according to the type of farming practiced, namely, industrialised commercial or subsistence farming. Organic farming has grown fairly rapidly in recent years, but this has been from a very small base and the sector comprises a miniscule proportion of commercial farms in South Africa (Niemeyer and Lombard 2003).

The industrialised, commercial agricultural system in South Africa is highly dependent on fossil fuel energy at every stage of the value chain, from primary

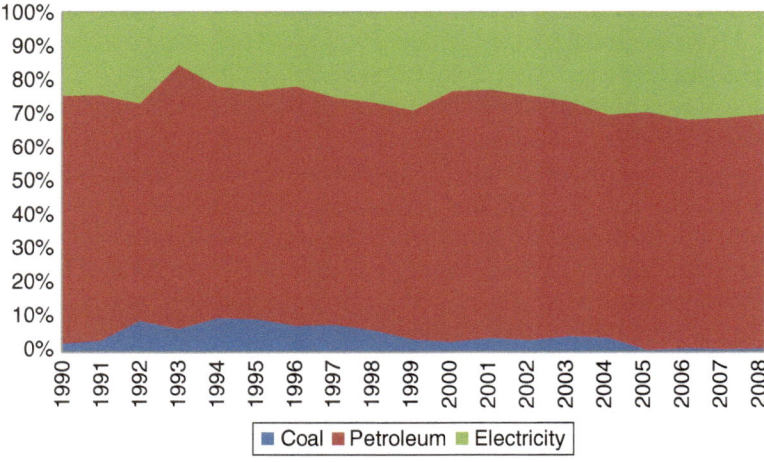

Fig. 4.1 Energy consumption in agriculture, 1990–2008. *Source*: IEA (2013)

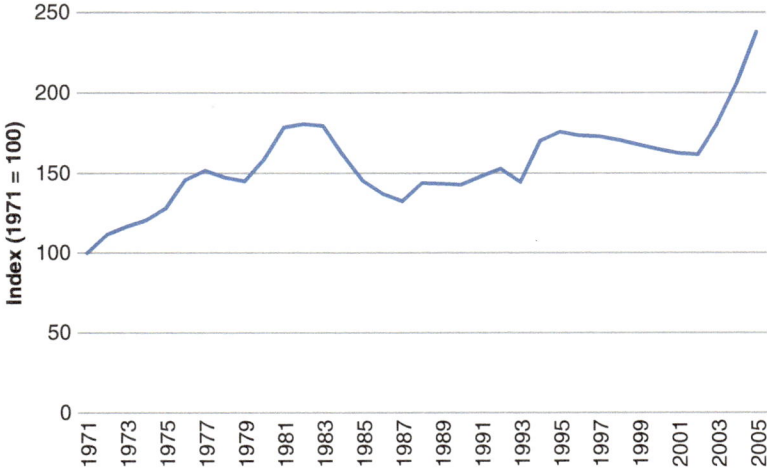

Fig. 4.2 Capital–labour ratio in agriculture, 1971–2005. *Source*: Author's calculations based on DAFF (2012) and SARB (2013). *Note*: The employment series from DAFF (2012) had several missing values, which were replaced with linearly interpolated figures

production on farms to processing in factories and to wholesale and retail distribution. At the production stage, this energy intensity results primarily from the extensive use of liquid petroleum fuels—especially diesel—to power farm vehicles and machinery such as tractors, planters, and harvesters. Electricity is also consumed to power irrigation systems and other machinery, including refrigerators. As Fig. 4.2 shows, the relative capital intensity (as measured by the capital/labour ratio) of commercial agriculture has increased considerably over the past several decades as farmers have progressively replaced human labour with machinery.

In addition to the direct use of fossil fuels, commercial agriculture consumes significant quantities of energy indirectly in the form of fertilisers and pesticides, whose manufacture involves the use of natural gas (or gasified coal) and oil, respectively. Intermediate input costs as a proportion of gross income in the agricultural sector have generally risen since the 1970s, with the exception of fertiliser costs (see Fig. 4.3). Most noticeable is the rising trend in the proportionate cost of fuel, from a low of 4.3 % in 1988 to a high of 9.3 % in 2008. Total input costs rose from an average of 33 % of gross income in the 1970s to 60 % in 2010 (DAFF 2012).

Traditional subsistence farming is generally much less dependent on oil than commercial farming, for several reasons. First, subsistence production is small scale and labour intensive rather than large scale and capital intensive (mechanised) and therefore uses little or no petroleum fuel directly. Second, traditional farming has at least until recently been mostly organic, i.e. farmers make little use of chemical fertilisers and pesticides derived from fossil fuels (Modi 2003: 676). Nevertheless, some subsistence farmers may rely to an extent on purchases of fertilisers, seeds, and other inputs whose prices may be affected by oil prices. Third, most smallholder produce is consumed locally rather than being transported to distant markets.

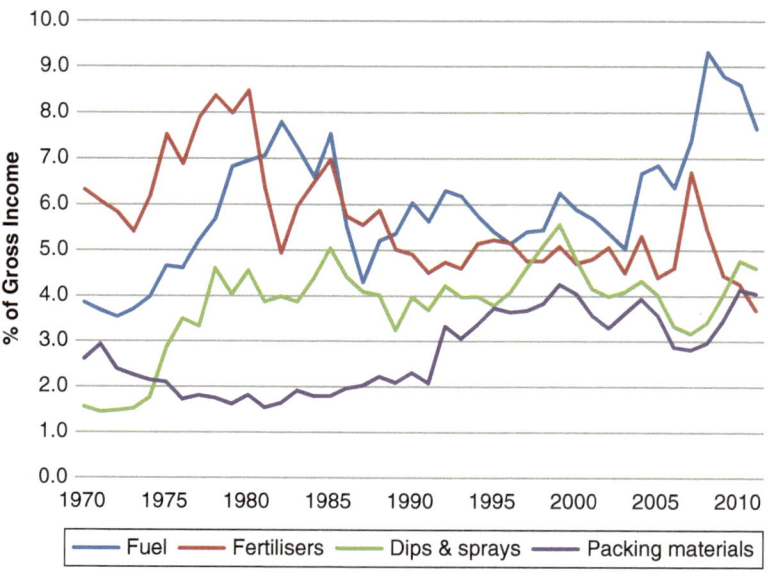

Fig. 4.3 Input costs as a percentage of gross income in agriculture, 1971–2012. *Source*: DAFF (2012). *Note*: Figures are for the 12 months ending in June of that year

4.1.1 National Food Security

The dependence of agriculture on oil has significant implications for national food security. According to the Food and Agricultural Organisation, "Food security exists when all people, at all times, have physical and economic access to sufficient, safe and nutritious food to meet their dietary needs and food preferences for an active and healthy life" (Hendriks and Msaki 2009: 184). Food security can be analysed on different scales of aggregation. The remainder of this subsection considers food security at the national level, while household-level food security is discussed in Sect. 6.2. At the national level, food security has two determinants: (1) the capacity of the country to be self-sufficient in food production and (2) the ability of the country to afford food imports where necessary or desirable.

South Africa has the capacity to be self-sufficient (i.e. domestic production exceeds consumption) in most agricultural products (GCIS 2009). On average, the country produces sufficient quantities of the main staple crop, maize, to meet domestic consumption (see Fig. 4.4). Historical exceptions to this have largely been the result of droughts. However, South Africa does rely on imports for some significant agricultural products. Major agricultural net imports include rice, wheat, poultry, and vegetable oils (DAFF 2013). Approximately one-third to one-half of the country's wheat requirement is imported, although this is partly because imports are cheaper than domestic production on marginal lands and there is no import tariff

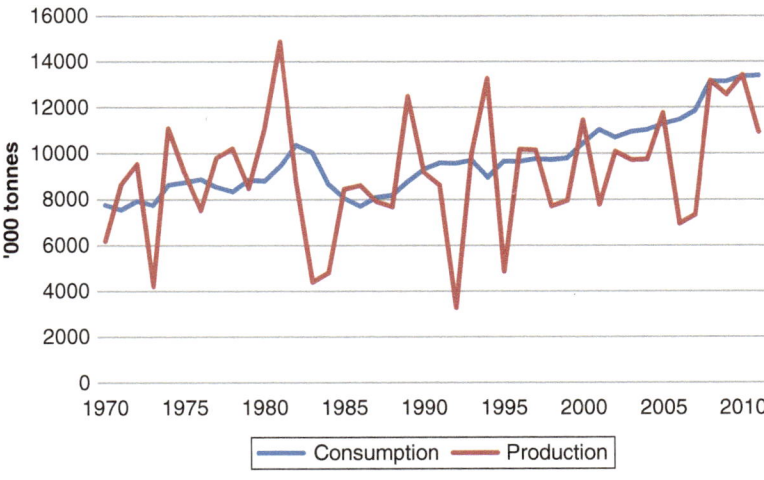

Fig. 4.4 Maize production and consumption in South Africa, 1970–2011. *Source*: DAFF (2012)

protection. All of the rice consumed domestically is imported, but rice is generally consumed by the wealthy minority and does not represent a broad staple food. Agricultural imports, mostly of processed foods, grew more rapidly than exports in the early 2000s so that by 2008 South Africa had become a net food importer in value terms for the first time (DAFF 2012).

Continued national food self-sufficiency clearly depends on access to affordable, quality inputs (such as fertilisers, pesticides, and machinery) for agricultural production. South Africa became a net importer of fertilisers in the 2000s (FAO 2005: 19). Domestic fertiliser prices are influenced heavily by prevailing international prices, the rand–dollar exchange rate, and freight costs (FAO 2005: 28) and are therefore susceptible to rising oil prices both directly (through higher transport costs) and indirectly (through the impact of oil prices on the exchange rate and international prices). Fertilisers are mostly delivered to farms by road and rarely by rail (FAO 2005: 30), further entrenching dependence on oil. The majority of farming equipment, such as tractors and harvesters, is imported, and therefore farmers face the risk of rising international prices and/or a depreciating exchange rate. All this means that food independence would be a difficult goal to achieve.

The second important determinant of national food security is South Africa's capacity to import food products. This in turn depends on international food prices as well as the strength of the domestic economy, in particular the balance of payments and the level of the exchange rate (see Chap. 5). These aspects of the macroeconomy are likely to come under pressure from any future oil price shocks. International food prices have risen dramatically over the past decade and are under threat from unstable weather patterns, rising demand due to population growth and increasing affluence in the developing world, and rising input costs (Brown 2012).

4.2   Impact of Oil Shocks on Agriculture

Rising fuel prices will raise direct input costs for fuel and chemical products that use oil (and oil substitutes) in their manufacture, including pesticides, fertilisers, and packing materials (see Fig. 4.5). In addition, rising transport costs will add to the prices of chemical inputs and raise the costs of transporting produce to food processors, wholesalers, and markets. If production costs rise faster than sales prices, then agricultural output will decline. Given the highly concentrated nature of the food processing and retail sectors in South Africa (Mather 2005), individual farmers are not able to pass on all cost increases to consumers, which exposes farmers to possible bankruptcy if costs rise too much. For exporters, higher world commodity prices could offset higher input costs, although higher transport costs might dampen foreign demand. Profitability would depend greatly on exchange rate movements. The short-term volatility in oil prices will create a great deal of uncertainty for farmers, who will face difficult choices about whether to plant crops and which crops to plant. Persistent higher oil prices and shortages of oil might encourage farmers to revert to more labour-intensive and organic methods of production that rely less on petroleum-based fuels and pesticides.

Any physical shortage of liquid fuels arising in rural areas would compromise the production of agricultural commodities. Since key farming operations, such as planting and harvesting, are highly time dependent, fuel shortages at such critical times could be devastating to output. Fuel shortages would also curtail the distribution of farming products to processing facilities and markets in towns and cities. The likelihood of fuel shortages emerging in rural areas is greater than that in urban areas due to the location of South Africa's oil refineries in or near to just four major urban centres (Cape Town, Durban, Mossel Bay, and Sasolburg and Secunda near Johannesburg).

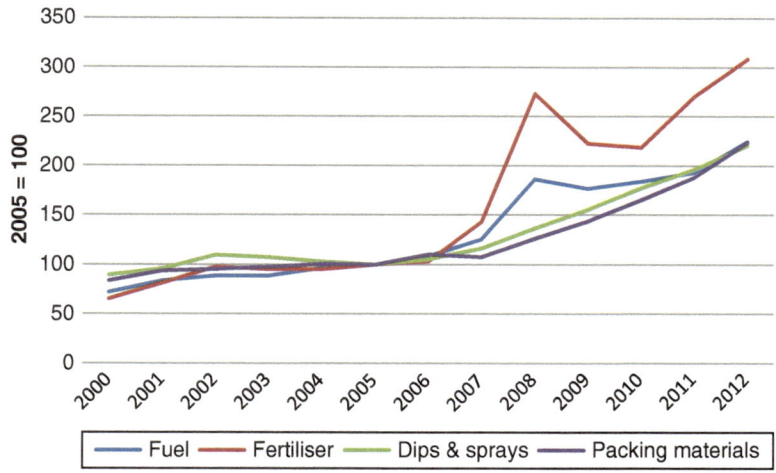

Fig. 4.5 Price indices of intermediate goods in agriculture, 2000–2012. *Source*: DAFF (2012)

4.3 Mitigating Peak Oil: Agroecological Farming and Re-localisation

As argued above, leaving the adjustment of commercial agriculture to increasing oil scarcity to occur through market forces could be very destructive to the sector, with potentially wide-scale bankruptcies and loss of output in the event of severe oil price or supply shocks. Thus there is a clear role for mitigation and adaptation strategies and policies. In the context of increasing world oil scarcity, there are several important objectives for national agricultural policy. The short- to medium-term goal should be to develop the resilience of agriculture to oil shocks. In particular, farmers should be protected from severe oil price shocks so that they do not become bankrupt, as this would compromise agricultural output and national food security for the long term. In addition, it is essential that the distribution of food products to markets be maintained. The longer term goals should be to systematically reduce oil dependency, to enhance food sovereignty, and most broadly to effect a transition to sustainable agriculture. Food sovereignty has been defined as "the right of each nation to maintain and develop their own capacity to produce foods that are crucial to national and community food security, respecting cultural diversity and diversity of production methods" (Pimbert 2008: 43). Sustainable agricultural systems may be defined as those that efficiently utilise environmental goods and services whilst preserving natural, human, social, physical, and financial capital (Hine et al. 2008: 6).

There are two core strategies for improving the resilience and sustainability of the commercial agriculture and food system. The first explores ways in which the use of petroleum fuels can be reduced in the food production system, together with specific policy measures and likely constraints. The second considers the distributional efficiency of food distribution systems and advocates a general reduction of distances between producers and consumers, that is, a re-localisation of the food system.

4.3.1 Reducing Oil Use in Agricultural Production

The main use of oil in agriculture in South Africa is in the form of diesel fuel for tractors, harvesters, and other farm machinery. However a wide range of other inputs are derived from crude or synthetic oil, including irrigation piping and fittings, pesticides, and packaging materials. In addition, rising crude oil prices will put upward pressure on the price of natural gas, which globally is the primary feedstock for synthetic nitrogen fertilisers. The adoption of agroecological farming practices presents the most complete way of reducing oil dependence in the agriculture sector.

Agroecology is "the science of applying ecological concepts and principles to the design, development and management of sustainable agricultural systems" (Pfeiffer 2006: 59). Practically, an agroecological approach involves "enhancing the habitat so that it promotes healthy plant growth, stresses pests, and encourages beneficial organisms while using labour and local resources more efficiently"

(Altieri 2009: 109). Organic agriculture may be defined as a "system of agricultural production that seeks to promote and enhance an ecosystem's health while minimizing adverse effects on natural resources" (Hine et al. 2008: 6). It aims to use locally available and natural materials as far as possible and thus inherently minimises the use of fossil fuel inputs, including oil. According to Pfeiffer (2006: 68), "organic farming is the most practical method of reducing fossil fuel input at the level of production". Clearly, the definitions of agroecological and "organic" farming are in practice very similar and will thus be used interchangeably hereafter.

The use of tractors can be reduced through the implementation of no-till or low-till agriculture. Conservation agriculture, which is based on a principle of minimal soil disturbance, reduces the need for tractor usage and therefore diesel fuel (FAO 2010). However, weed management becomes more challenging, which given the unsustainability of herbicides implies greater demand for labour (Giller et al. 2009). Another option is to utilise alternative energy sources for farm machinery. Biodiesel can be manufactured on a small, local scale from crops produced on a portion of a farmer's land. Solar-powered electric tractors have been designed that can be recharged from the grid or from tractor-mounted photovoltaic panels (Heckeroth 2009). Alternatively, draft animals can be substituted for tractors, bringing additional advantages of less compaction of soils and the generation of manure for fertilising. However, both biodiesel and animal power imply a reduced land area available for food production.

Another important aspect of organic agriculture is soil rehabilitation to restore soil fertility without inorganic fertilisers (Heinberg and Bomford 2009: 22). This can be achieved through appropriate crop rotation, incorporating nitrogen-fixing crops, and recycling of critical nutrients (including phosphorus) through the use of composting, animal manures, green manures (Pfeiffer 2006: 58), and even "humanure" (Greer 2009). Animal and human manures were the chief sources of fertiliser before the fossil fuel era, although yields were much lower. Using animal manure implies allocating more arable land to grazing and therefore less to growing food crops. Rehabilitating depleted soils takes several years of sustained effort (Heinberg and Bomford 2009: 22). In addition, according to the foregoing authors, oil-based pesticides could in principle be replaced with integrated pest management, utilising biopesticides, microbes, and natural pest control; intercropping to reduce losses to pests; and cover cropping to counteract weeds, although this may be difficult to achieve in practice.

Most commercially available seeds are produced through energy-intensive, centralised production and distribution systems, and seeds for some staple crops such as maize and soya include an increasing percentage that is genetically modified (GM), which typically are tied to chemical fertilisers and pesticides (Heinberg and Bomford 2009: 30). More preferable would be programmes to identify and distribute seeds of locally adapted, open-pollinated food crops and training in seed-saving techniques. Government could also sponsor the development of regional heirloom seed banks.

A shift to agroecological farming methods could bring additional benefits beyond reduced oil dependency and lower production costs (Hine et al. 2008). They can boost stocks of natural, social, human, physical, and financial capital in rural communities and thus have a lasting impact on food security and well-being.

Where organic farmers are able to produce surpluses for export, their products command premium prices and can therefore boost incomes and alleviate poverty amongst small farmers. Environmental benefits include reduced pollution (e.g. from pesticides, fertilisers, and greenhouse gas emissions) and enhanced soil fertility, water quality, and biodiversity.

A reversal of the historical trend towards integration of farms into larger units is a likely consequence of increasing oil scarcity since the energy-related costs of large-scale industrial farming methods will rise significantly. If this process is to unfold beneficially, it may need to be supported by government policies. Organic farming is more efficient when conducted on a small scale and is inherently more labour intensive than industrial agriculture and therefore has the potential to create numerous livelihood opportunities. Furthermore, small farms have "multiple functions which benefit both society and the biosphere" (Rosset 1999: 452). Potential benefits of small farms include greater diversity, improved natural resource management, community empowerment and responsibility, and development of the local economy through multiplier effects (Rosset 1999). Rosset also cites studies showing that the productivity of small farms, which usually have multiple crops, is generally greater than that of large farms when taking into account the per-hectare total yield (of all crops) as opposed to the yield from a single crop. A study comparing the performance of organic and conventional farming systems over a 22-year period in the American state of Pennsylvania found that organic farming had several environmental advantages (such as higher levels of organic matter and nitrogen in soils, better conservation of water, lower fossil fuel inputs, and reduced soil erosion) and that "[d]epending on the crop, soil, and weather conditions, organically managed crop yields on a per-ha basis can equal those from conventional agriculture, although it is likely that organic cash crops cannot be grown as frequently over time because of the dependence on cultural practices to supply nutrients and control pests" (Pimentel et al. 2005: 580). Traditional agriculture involved small-scale farms with high levels of genetic and biological diversity and reliance on local resources and knowledge, all of which made farming resilient to changing conditions (Altieri 2009: 103). Altieri (2009) suggests that this form of farming can provide a model for the future, as it does not rely on oil and chemicals and is more ecologically and socially sustainable than industrial agriculture.

Nevertheless, the transition to sustainable agriculture will involve several challenges. First, good-quality land and water are both scarce in South Africa. Second, the current dearth of farming skills implies that considerable time would be needed to train new farmers in agroecological methods. Third, the agroecological model of farming requires more land than industrial farming as part of the land has to be set aside for animals to produce manure for fertilising (Pfeiffer 2006: 63). Agroecological farming also has greater labour requirements, but this can be seen as an opportunity to create sustainable, if low paid, livelihoods. Fourth, farmers who convert from conventional to organic farming will face transition costs in the form of initially reduced output and revenues (to some extent offset by reduced input costs) and human capital investment costs (Hine et al. 2008: 34). Nevertheless, Hine et al. (2008: 11) state that while farms that convert from industrial to organic methods generally experience a decrease in yield initially, yields increase notably once the agroecosystem recovers. On the other hand, large-scale, mechanised farms are able to reap economies of scale and thus produce at lower prices, although some costs are externalised to the environment. In addition, fertilisers and pesticides have allowed the cultivation of more marginal soils, which might not be suitable for organic production. Thus total output could decline significantly when fossil fuel inputs are curtailed, as it did in the early 1980s after fertiliser subsidies were reduced (FAO 2005).

The successful diffusion of agroecological innovations requires specific policy support and institutions, rather than being side-lined in favour of high-input farming approaches (Altieri 2009; Hine et al. 2008). This includes the establishment of a clear policy and regulatory environment and greater priority for agroecological farming in scientific and research budgets. Moreover, various networks need to be strengthened, for example, those involving scientists, agricultural extension providers and farmers and connections between farmers, civil society organisations, and government departments.

In Cuba, the establishment of thousands of private cooperatives, managed and owned by farm workers, helped the transition to sustainable agriculture following a drastic decline in Cuba's oil imports after the collapse of the Soviet Union (Pfeiffer 2006). Cooperatives reward individual members for their productivity and yet offer the benefits of economies of scale. Through the Cooperatives Act of 2005, the South African Government has sought to promote the horizontal integration of emerging farmers to help improve their access to markets and to share resources (Lyne and Collins 2008). However, Lyne and Collins note that the historical experience of development-oriented cooperatives in South Africa has been very poor. They argue that the Cooperatives Act should be amended to allow for non-patron members of "emerging" cooperatives in order to facilitate the transfer of knowledge and capital from private agribusinesses.

Access to land and water are key to achieving food sovereignty and food security (Altieri 2009), while the transition to organic agriculture will require an increase in the number of farmers and farm workers due to its small-scale and labour-intensive nature. Thus effective land reform that preserves agricultural output will be an essential component of reducing oil dependency and vulnerability. An argument has been made that land reform should be restricted to apply mainly to small-scale commercial farmers, since they contribute a relatively small share of agricultural output, while existing large-scale farmers produce the bulk of agricultural value and are thus important for maintaining national food security (Vink and van Rooyen 2009: 34). However, this perspective assumes the continued viability of large-scale, oil-intensive farming, which may become increasingly unviable as fossil fuel input costs rise. Crucially, land reform must be accompanied by adequate training and skill acquisition for the new farmers; otherwise, the result can be unsuccessful and detrimental to food production and food security. The government's on-going land reform process must be sensitive to the additional pressures placed on the agriculture sector by growing oil scarcity.

Various other farmer support services can also assist the move to sustainable agriculture. Financial support may be needed to enable new farmers to start up. This could be provided in the form of subsidies or low-interest credit. Any subsidies provided to farmers should be targeted to those who are progressively implementing more sustainable (e.g. organic) agricultural methods; otherwise, unsustainable practices and petrochemical dependency will be perpetuated. While politically unpopular at present given the racially skewed patterns of land ownership, farming subsidies would reflect the national priority of agriculture and food security in ensuring social welfare and cohesion. In addition, the knowledge-intensive nature of organic farming means that learning capabilities and cooperation need to be enhanced, for example, through investments in local social capital (Hine et al. 2008). The dissemination of knowledge, skills, and training should be "targeted at those who need it most, especially farmers in remote rural areas" (Vink and van Rooyen 2009: 36). Other forms of support include "assistance in accessing commercial supply chains", including support for collective action and access to alternative markets not dominated by large processors and supermarket chains (Vink and van Rooyen 2009: 36).

4.3.2 Distributional Efficiency and Localisation

The vulnerabilities of the agricultural and food distribution systems to rising oil prices and fuel shortages can be addressed in two main ways. In the first instance, targeted interventions should be made to enhance logistical efficiencies of supply chains operating between input suppliers and farms and between farms and consumers (Vink and van Rooyen 2009: 37). Resilience to fuel supply shortages can be increased by building in redundancies and increasing inventories (Heinberg and Bomford 2009: 33). Second, given the overwhelming dependence of the current distribution systems on road-based transport infrastructure, for the longer term a fundamentally new orientation is required: a re-localisation and decentralisation of food economies that reduces distances been producers and consumers. Re-localisation implies that production of a greater proportion of necessary foodstuffs occurs locally, while longer distance trade is reserved mainly for luxury items (Heinberg and Bomford 2009: 15). In order to promote food security, each area should produce regionally adapted staple crops as far as possible. For food processing, re-localisation would involve the establishment of smaller scale facilities rather than large, centralised processing plants.

Localisation of food economies may bring several benefits beyond fuel saving. First, food availability is more seasonal and often of better quality, with less processing required (Heinberg and Bomford 2009: 16). Second, localisation has been shown to build community networks and promote resilience (McKibben 2007). Third, local food economies "tend to promote greater sustainability by shifting the decision making around the food system back to the communities in which they are embedded" (Schulschenk 2010: 122). Fourth, greater prioritisation of import replacement and production for local needs would reduce exposure to fluctuations in foreign demand and exchange rates while promoting diversity and strengthening the regional economy (Shuman 1997). On the other hand, local food economies may be limited in their capacity to meet diverse nutritional needs of some communities and constrained by environmental conditions such as extreme climates or degraded ecosystems (Schulschenk 2010: 59–61).

Localisation will require a major redirection of national agricultural policy, which currently favours an export-led path. At the micro level, government can help to establish localised agricultural markets and to promote farmers' markets, for example, by making public spaces available in urban areas and rural towns. Such local markets help to reduce the power of giant wholesalers and retailers and strengthen community bonds (Pfeiffer 2006). Government can also support "buy local" campaigns and insist that government institutions source a minimum portion of their food needs locally (Heinberg and Bomford 2009: 17). Building local food economies initially may depend on civil society participation to be competitive with the current food system, although over time rising oil prices will assist the process (Schulschenk 2010: 122).

The development of urban agriculture (UA) represents a specific form of localisation with significant opportunities to foster the resilience of urban communities (Hopkins 2000). Local governments can promote urban food production by allocating underutilised land for food gardens and leasing allotments to residents. Local by-laws can foster rooftop and backyard food gardens, and laws could be enacted to require urban agricultural produce to be organic, as in Cuba (Pfeiffer 2006: 61). Municipalities can either organise their waste systems to process food waste into compost or biogas (Heinberg and Bomford 2009: 17) or incentivise residents to do so themselves. The National Department of Agriculture could widen its extension services to cover urban townships and informal settlements (Thornton 2008). Yields from bio-intensive UA have been shown to be higher than those of conventional farming in the United States (Hopkins 2000: 206). Also in the United States, Will Allen's "Growing Power" program has provided a practical model for successful UA (Allen and Wilson 2012). Nevertheless, it is recognised that the development of UA in South Africa would face significant challenges and constraints, such as the availability of suitable land, water, and organic materials for composting as well as security. Evidence suggests that cultural attitudes about the "backwardness" of food production and the availability of social welfare grants can also hinder the growth of UA (Thornton 2008).

4.4 Conclusion

The vast majority of agricultural produce in South Africa comes from highly oil-intensive commercial farming, which is exposed to rising input costs for fuel, fertilisers, pesticides, and packaging, all of which are affected by international oil prices. Fuel shortages are potentially catastrophic at critical times such as during planting and harvesting. Moreover, the distribution of food products to consumers relies heavily on road-based transport and therefore on liquid petroleum fuels. Food security concerns represent a strong case for the commercial agriculture sector to receive government support to cope with oil price shocks and potential fuel scarcity in the short to medium term. This support could be in the form of subsidies and priority fuel allocations in times of shortage. Politically, however, farming subsidies would encounter much opposition given the historical inequities in the racial distribution of land ownership; the ANC government in fact moved in the opposite direction after taking power in 1994, by changing many policies that historically supported white farmers.

If the government should at some point determine that fuel price shocks or shortages pose a sufficient threat to food security (and therefore social stability) to warrant financial support for farmers, such support should ideally be linked to a programme aimed to gradually phase out the use of petroleum products as farming inputs and undertake a gradual transition to sustainable, agroecological farming practices. This transition will require institutions that support both existing and emerging farmers with appropriate knowledge and skill acquisition. Farmers will need to invest in renewable energy and adopt farming practices that rely less on mechanisation, such as conservation tillage. Efforts to mitigate declining oil availability need to take cognisance of various additional vulnerabilities of the agriculture system, such as low and declining soil quality, water scarcity and droughts, and the likely impacts of climate change. It should also be managed as a phased transition, since "[r]emoving fossil fuels from the food system too quickly, before alternative systems are in place, would be catastrophic" (Heinberg and Bomford 2009: 12). Nevertheless, serious questions remain about the potential of agroecological farming methods to feed South Africa's population with greatly reduced fossil fuel inputs. Furthermore, there is the additional challenge of transporting food to the towns and cities where over 60 % of the South African population reside. This calls for a re-localisation of agriculture, including the development of urban agriculture, to reduce the distances that food products must travel to reach consumers. An additional strategy for the long term would be to facilitate the development of smallholder farming through appropriate training and support. This in turn has implications for land ownership patterns, a thorny issue that the government has been grappling with—with limited success—for nearly two decades.

All in all, there are no easy solutions to the challenge of reducing the oil dependency of agriculture and food systems in South Africa. The next chapter broadens the scope of analysis to consider the economy-wide implications of peak oil and how policymakers could respond proactively to ameliorate negative macroeconomic impacts.

Chapter 5
Economy

By developing country standards, South Africa's economy is reasonably well diversified across the range of sectors. This diversity is partly a legacy of the country's isolationist past in the apartheid era, when import substitution was a survival strategy rather than a policy choice. The evolution of the broad structure of the South African economy between 1970 and 2012 is represented in Fig. 5.1, which shows the relative contributions to gross value added (GVA) by the 9 one-digit economic sectors. The most significant structural shift has been the decline in the relative importance of mining, from over 20 % of GVA in 1970 to 5.5 % in 2012. Throughout the period agriculture was the second smallest sector, accounting for just 2.4 % of GVA in 2012. The largest gains were in transport (almost doubling in relative size from 5.4 % in 1970 to 10.1 % in 2012) and financial services (15–24 %), the latter demonstrating the trend towards financialisation of the economy. Other sectoral shares have not changed notably. The "community, social, and personal services" sector includes government services such as education and healthcare. In 2012 the tertiary sectors together contributed 69.5 % of GVA, secondary sectors 22.6 %, and primary sectors (agriculture and mining) just 7.9 %, which is broadly in line with other upper-middle-income countries.

The South African Government has articulated a conception of a "Second Economy," existing alongside the formal (first) economy and reflecting endemic "structural inequalities, disadvantage, and marginalisation" in society (The Presidency 2008: 40). To a large extent this is a legacy of apartheid-era policies which severely restricted land ownership and educational opportunities among black South Africans. However, unemployment rates have soared since the democratic transition in 1994, inter alia because of the relaxation of trade barriers and the consequent erosion of globally uncompetitive industries such as clothing and textiles, a persistent skills constraint, and continued mechanisation of agriculture.

The diversity of the formal economy bodes well in terms of its resilience to the anticipated impacts of higher oil prices. Nevertheless, history suggests that oil shocks could have substantial negative impacts on the macroeconomy. This chapter begins with an analysis of the economy's energy intensity and oil dependency.

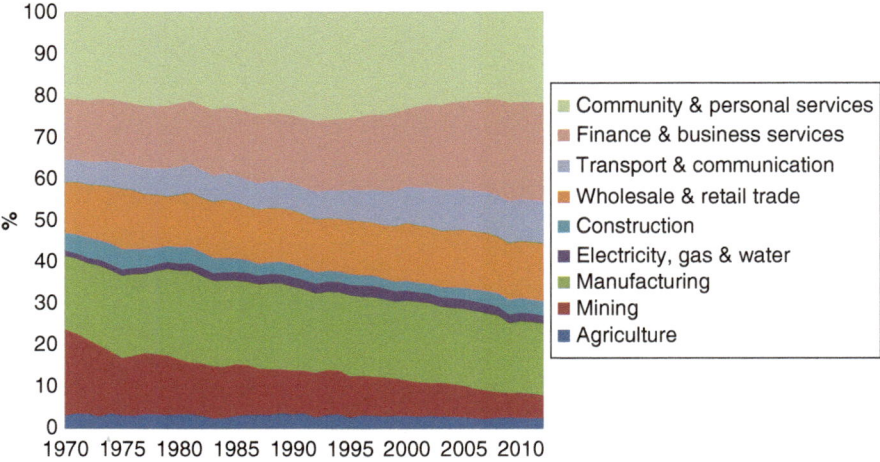

Fig. 5.1 Sectoral shares of real gross value added, 1970–2012. *Source*: SARB (2013)

The second section highlights the main macroeconomic strengths and vulnerabilities in the face of oil shocks, while the third section discusses their likely macroeconomic impacts. The fourth section details mitigation strategies in the areas of fiscal, monetary, industrial, trade, and labour market policy.

5.1 Oil Dependence of the Economy

The South African economy is characterised by high *energy intensity* (i.e. energy used per dollar of GDP), especially in its mining and manufacturing sectors. The country's abundant coal reserves have historically been exploited to generate amongst the cheapest electricity in the world, which supported both the extensive mining industry as well as energy-intensive mineral beneficiation. South Africa's energy intensity, measured as the ratio of total primary energy supply to real GDP, rose by 40 % between the 1970s and 1980s but since the mid-1990s has declined by approximately 20 % (see Fig. 5.2). However, when measured by the ratio of total final energy consumption (i.e. consumption of energy carriers such as electricity, petroleum products, coal, and biomass) to real GDP, energy intensity has gradually fallen by a cumulative 20 % since the late 1970s. This partly reflected a change in the composition of the economy away from mining and manufacturing to services, which are less energy intensive, but to some extent was a result of improvements in energy efficiency. The divergence between the two measures of energy intensity was apparently driven mainly by the substitution of domestically produced synthetic liquid fuels derived from coal for imported oil (the rise in the TPES energy efficiency index in the early 1980s coincides with the construction of the second and third Sasol CTL plants). In other words, a lower quality energy source (coal) has been substituted for a higher quality source (oil).

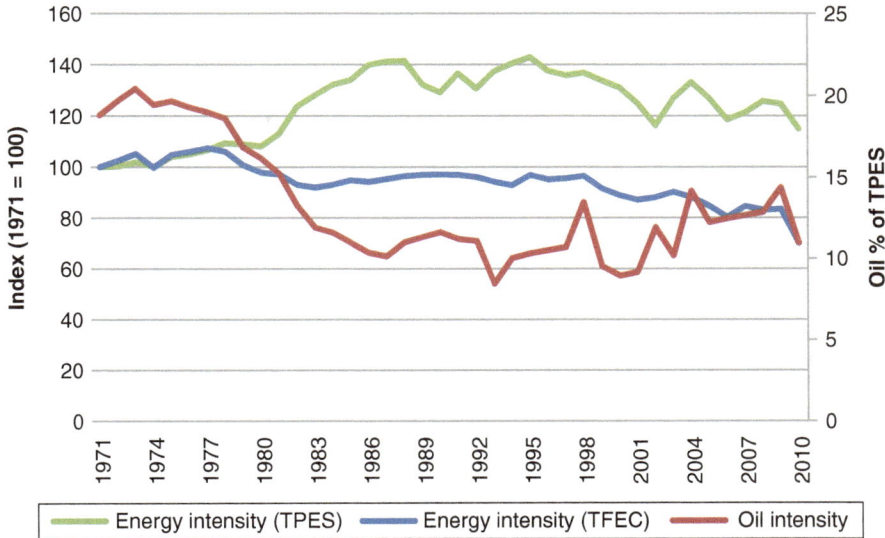

Fig. 5.2 South Africa's energy intensity and oil resource intensity, 1971–2010. *Source*: Author's calculations based on IEA (2013); SARB (2013). *Note*: The energy intensity indexes are derived from the ratios of total primary energy supply (TPES) and total final energy consumption (TFEC, as defined by the IEA) to real GDP, respectively; oil intensity is the percentage share of oil in total primary energy supply (TPES)

South Africa's *oil resource dependence* is low relative to many other developing countries, partly as a result of the heavy use of domestic coal reserves. According to IEA figures, oil in 2010 constituted approximately 11 % of South Africa's primary energy supply, down from around 20 % in the early 1970s (see Fig. 5.2 above). Again, the decrease in the early 1980s was mainly attributable to the expansion of Sasol's CTL production. The slight rise in oil intensity in recent years has been driven mainly by increasing private vehicle ownership and road freight movement but also by the use by Eskom of diesel to fuel its open-cycle gas turbine power plants.

Because South Africa's domestic oil reserves are very limited, the country has a very high degree of *oil import dependence*: over 95 % of crude oil requirements are imported. However, thanks to a strong domestic liquid fuels industry (Sasol's coal-to-liquid and PetroSA's gas-to-liquid facilities), only 70 % of the country's *liquid fuel* requirements are met by imported oil. On the other hand, the synthetic liquid fuels produced by Sasol and PetroSA are currently priced on an import parity basis. If this does not change, then consumers are just as vulnerable to oil *price* shocks, even though synfuels provide a partial buffer for oil *supply* shocks.

The nominal value of South Africa's crude and refined oil imports rose rapidly between 2004 and 2008 (see Fig. 5.3), thanks to a combination of rising consumption (driven by economic growth and an expanding population) as well as a steadily rising oil price. In 2008 the country spent nearly R138 billion, or 6 % of GDP, on oil imports, which represented the single largest import item on the balance of payments. In 2009, as a result of the recession and lower oil price, oil imports fell to

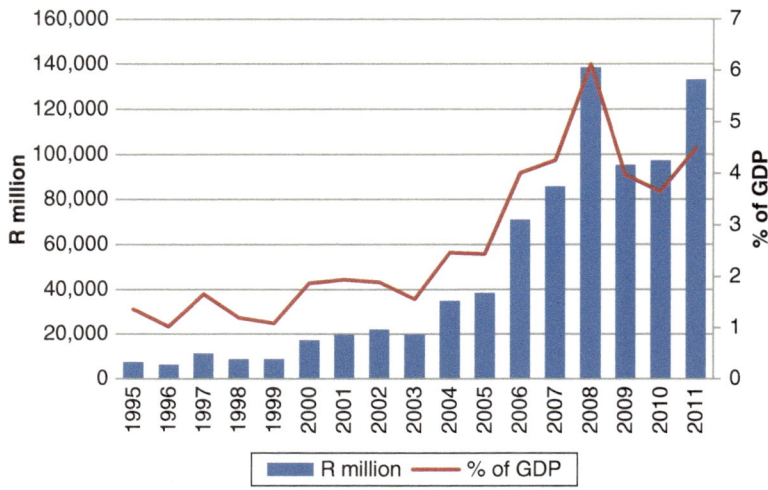

Fig. 5.3 Net crude and refined oil imports, 1994–2011. *Source*: DTI (2012) and SARB (2013)

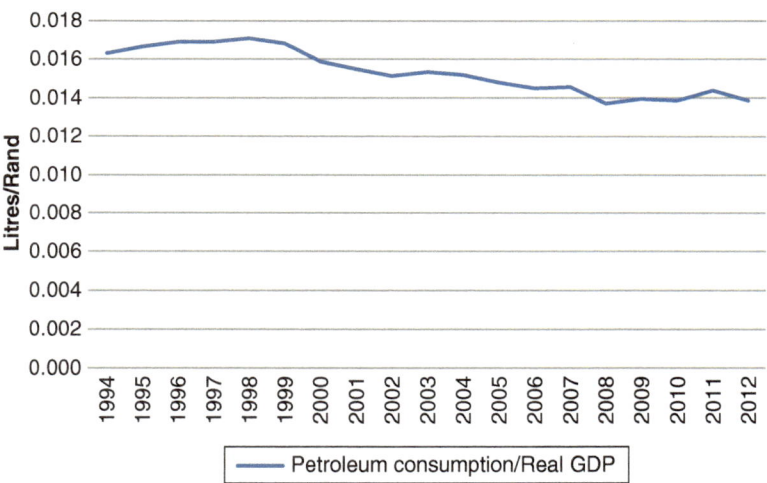

Fig. 5.4 Petroleum intensity of the economy, 1994–2012. *Source*: Author's calculations based on SAPIA (2013) and SARB (2013)

R95 billion (4 % of GDP). In comparison, net coal exports in 2009 amounted to R33 billion (1.4 % of GDP). By 2011 expenditure on oil imports had begun to rise again, reaching 4.5 % of GDP.

A final measure of the oil dependence of the South African economy is provided by the ratio of petroleum product consumption to real GDP. Figure 5.4 shows that this measure of petroleum dependency increased slightly from 1994 (by 4.5 %) to reach a peak in 1998, after which it declined nearly monotonically by an average of 1.4 % per annum and a cumulative 19 percentage points by 2012. This is an illustration of *relative resource decoupling,* defined as "reducing the rate of use of (primary) resources per unit of economic activity" (Fischer-Kowalski and Swilling 2011: 4); although absolute consumption of petroleum products rose between 1998 and 2012, consumption relative to real GDP fell considerably. This relative decoupling can partly be explained by the expansion of the financial services sector, which has very low petroleum intensity. Nevertheless, this decoupling achievement bodes well for the potential to reduce future petroleum consumption while attenuating negative impacts on economic activity.

5.2   Macroeconomic Vulnerabilities

The major vulnerabilities of the South African macroeconomy in the face of international oil price shocks, as of 2013, were as follows:

- The current account deficit, which reached 7.2 % of GDP in 2008 when oil prices spiked and stood at 6.3 % of GDP in 2012, represents a significant risk in terms of potential currency volatility, speculative attacks, and exchange rate depreciation.
- The financial account of the balance of payments has relied heavily on short-term portfolio inflows and is highly vulnerable to sudden capital flight, which therefore poses a significant currency risk.
- There has been a long-term decline in output of gold from South African mines since the early 1980s, mainly as a result of the depletion of ores, such that by 2012 gold exports accounted for just 8 % of total exports and 2.3 % of GDP (see Fig. 5.5 below). Thus while gold and other mineral exports can still be expected to provide some level of shock absorption for future oil price hikes, as they did in the past, this potential is substantially reduced relative to the 1970s.

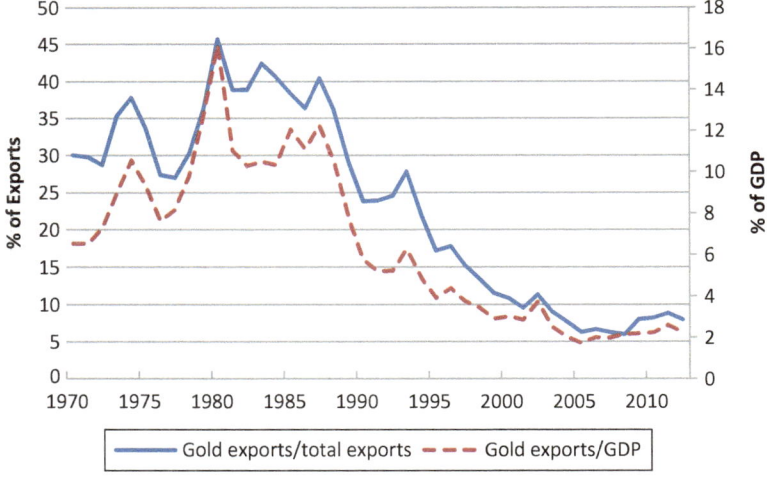

Fig. 5.5 Ratios of net gold exports to total exports and GDP, 1970–2012. *Source*: SARB (2013) and author's calculations

- While the National Treasury had achieved a fiscal surplus in 2006 and 2007, the fiscal deficit rose to 5 % of GDP in 2009 as a consequence of the Global Financial Crisis and ensuing recession in SA and has remained above 4 % in subsequent years. Consequently, the ratio of public debt to GDP rose from a historical low of 27 % in 2008 to 42.3 % in 2012—although this was still considerably below the peak of 50 % recorded in 1995 (SARB 2013). The large fiscal deficit and debt leave little fiscal space for responding to future shocks.
- Total foreign debt as a percentage of GDP rose from a low of 19.7 % in 2005 to a two-decade high of 35.8 % in 2012 (SARB 2013).
- The ratio of household debt to disposable income peaked at 82.4 % in 2008 before declining somewhat to 75.6 % in 2012 (SARB 2013). Nevertheless, this high level of debt exposes consumers to fallout from oil price shocks such as higher costs of living, higher interest rates, and falling real incomes.
- A major vulnerability is the unemployment rate, whose official figure stood at 25.6 % in the second quarter of 2013 (StatsSA 2013). The so-called broad rate of unemployment, which includes discouraged worker seekers (i.e. those who wanted to work but did not actively search for jobs in the week preceding the survey), was over 35 %.
- The headline consumer inflation rate crept above the South African Reserve Bank's (SARB) inflation target range of 3–6 % in July 2013 (StatsSA 2013). Since the Reserve Bank reduced its benchmark repurchase rate to a historical low of 5 % in June 2012, it has been caught between a rock of rising inflation and a hard place of slowing economic growth—i.e. a mild case of stagflation. This environment makes it difficult for the monetary policy authorities to respond to externally driven inflationary shocks or recessionary conditions.

5.3   Economic Impact of Oil Shocks

The direct impacts of oil price shocks occur via higher fuel prices and have reverberations on several important macroeconomic variables. South Africa is a price taker on the international oil market. Domestically, the downstream liquid fuels industry is subject to extensive government regulation. Prices of petroleum fuels (petrol, diesel, paraffin, and LPG) are administered by the state, which imposes various levies and taxes and determines retail and wholesale margins, over and above a "basic fuel price". The basic fuel price is determined by an import parity pricing formula which depends on the international spot price of refined oil. Sasol's and PetroSA's synthetic liquid fuels (converted from coal and gas, respectively) are accorded the same status in the domestic market as fuels that are refined from imported crude oil. The basic fuel price is influenced by two primary factors: the dollar price of crude oil traded on international markets and the rand/dollar exchange rate. Volatility in both of these variables has historically had a significant impact on the rand-denominated price of oil (see Fig. 5.6). Notably, the oil price shocks of 1973 and 1979 were somewhat muted in rand terms, thanks to the relative strength

Fig. 5.6 Real oil price in 2012 dollars and rands, 1970–2012. *Source*: IMF (2013) and own calculations

of the rand at those times (supported by a high gold price). The rand oil price in 2008 was almost double that of 1979 in real (inflation-adjusted) terms.

I undertook a comprehensive review of historical experience and various empirical estimates of the impact of oil price shocks on South Africa and found the results of numerous studies to be broadly consistent with one another (Wakeford 2012). In general, crude oil price shocks resulted in a depreciation of the exchange rate; an initial boost for some export commodities, such as coal and uranium, but a decline in certain other exports after a time lag; higher rates of producer and consumer price inflation; lower (or negative) growth in real GDP; falling employment and real wages; and greater poverty and inequality. The sectors most adversely affected include agriculture, light manufacturing, and private services, while the sectors benefitting most in relative terms were domestic synfuels, electricity, and coal and gold mining. The oil price shocks of the 1970s and 2008 can arguably be identified as at least major contributing factors to the ensuing recessions (see Fig. 5.7), mainly via their impact on global demand for South Africa's export commodities. However, the US experience of a tight correlation between oil price spikes and recessions (Hamilton 2009) has not been replicated in South Africa. The economic impact of world oil price shocks appeared to occur with a time lag of around 2–3 years in the 1970s but about 1 year in the case of the 2007/2008 price spike. It is notable that in the latter period (1) gold provided a much smaller buffer and (2) South Africa's economy was much more integrated into the global economy, which could have shortened the impact time lag. It seems that South Africa's vulnerability to global oil shocks has increased over time as a result of the country's reintegration to the world economy.

In the medium-term future, it can be expected that South Africa will suffer similar types of negative consequences from oil price and supply shocks as it did in 2008–2009. Given that both the global and South African economies are in a much

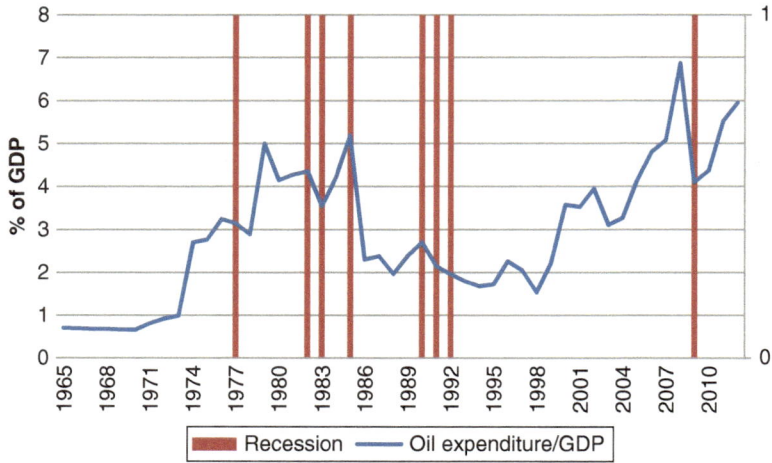

Fig. 5.7 Oil expenditures and recessions in South Africa, 1965–2012. *Source*: Author's calculations based on BP (2013); IMF (2013); SARB (2013)

weaker state in 2013 than they were immediately prior to the 2008 oil shock, future oil shocks might have even more severe socioeconomic effects.

5.4   Mitigating the Economic Impacts of Peak Oil

Macroeconomic policy is a vital component of a strategy for mitigating the impact of global oil depletion on South Africa's socioeconomic system. In this context, the short- to medium-term goal of macroeconomic policy should be to promote resilience to oil price shocks and physical oil shortages. The longer term objective should be to decouple economic development from oil consumption and to improve the sustainability of the macroeconomy. On a pragmatic level, macroeconomic policies and planning should be formulated with due cognisance of peak oil and its probable effects. This implies recognition of the likely reversal of globalisation in the trade of physical goods and some services (i.e. a re-localisation of economic production and consumption) and possible long-term global economic stagnation or even contraction.

5.4.1   Fiscal Policy

The key challenge for fiscal policy presented by the impacts of peak oil consists in maintaining fiscal sustainability. The budget deficit and therefore the level of public debt increased significantly during the 2009 recession and has remained stubbornly high since then as a result of growing expenditures and constrained revenues. There is a significant risk that the onset of declining global oil production will result in worldwide economic contraction and a recession in South Africa. This would increase social pressure on government relief spending, but at the same time tax revenues would shrink as business profits and consumer spending fall. Resorting to increased borrowing in the short term would be self-defeating in the longer run since the economic shocks will in all likelihood recur and possibly worsen. Thus in order to avoid larger budget deficits and an increasing borrowing requirement, which over time could take the country towards the level of a debt trap, the National Treasury will have to ensure that spending is in line with tax revenues. In addition, the Treasury should set an example for the broader economy by maintaining public sector wage restraint so as to avoid a potential wage–price spiral, as occurred in many countries following the oil shocks of the 1970s. Within a stance of fiscal prudence, there is considerable scope for expenditure switching in order to mitigate the impacts of global oil depletion. In general, given the constraints on government revenues noted above, mitigation strategies should wherever possible aim to be revenue neutral. For example, fiscal incentives can be realigned to encourage firms and consumers to invest in more sustainable, less oil-dependent capital stock (durable and semi-durable goods).

On the expenditure side of the budget, the priority should be to eliminate spending on infrastructure that would be imprudent in the context of high oil prices and fuel shortages. As discussed in earlier chapters, government spending should be prioritised on more sustainable infrastructure and services such as energy-efficient electrified public transport, renewable energy (provided the energy return on investment is sufficiently high), and telecommunication networks that reduce the need for physical travel. In addition, government procurement policy should aim to stimulate local production, as mentioned in the New Growth Path (EDD 2010), for example, in agriculture and light manufacturing (Fofana et al. 2008: 13). In the context of severe budgetary constraints, subsidies would in general be unsustainable, except to the extent that they provide an initial boost to green industries and activities that need initial assistance to achieve economies of scale in production. Subsidies to the automotive industry, which totalled R51 billion in the 3 years to 2011 (Donnelly 2011), should be reduced and tied to vehicle efficiency improvements and the production of BEVs and PHVs, if allocated at all. Tax rebates could be introduced for energy efficiency as this could save money in the long run (although the rebound effect can work against actual fuel savings). Any broad liquid fuel subsidy would be fiscally unsustainable and would encourage continued petroleum dependency. However, the fuel price stabilisation fund could be used to limit extreme short-term fuel price fluctuations (both increases and decreases), although the degree of uncertainty surrounding future prices would make its administration difficult. While there is a strong temptation for government to protect vulnerable sectors with temporary subsidies, these would in most cases be unsustainable in the face of repeated oil shocks. The exceptions would be industries that in the long run would benefit from the import substituting impact of higher transport costs and would promote greater sustainability.

On the revenue side of the budget, various taxes can be used to boost the greening of the economy. The Minister of Finance announced in his 2013 Budget Speech that a carbon tax would be phased in from 2015, following the earlier introduction of a very modest tax on coal-fired electricity and another on vehicle emissions. State levies already make up nearly a third of petrol and diesel retail prices. While carbon taxes can be effective in shifting incentives for energy efficiency and "clean energy", they may become increasingly unpopular as oil and other fossil fuel prices continue their upward trajectory. Furthermore, blanket carbon taxes provide no guarantee of fairness in the distribution of an increasingly scarce resource; that is, while richer consumers will be able to absorb the increased cost of fuels, poorer households may not. There might, however, be a case for gradual, predictable increases in liquid fuel taxes in the short term (before the next major oil price spike) in order to incentivise fuel economy initiatives while generating greater revenues for public transit infrastructure (Litman 2008). The Fuel Equalisation Fund, which historically was used to smooth out fluctuations in petroleum fuel prices, could be used again for this purpose. The fact that consumption of petroleum fuels is very likely to decline in the future implies that the National Treasury should plan for falling fuel tax revenues, which amounted to approximately R48 billion (7.2 % of total government revenue) in 2010. The Treasury will therefore need to formulate alternative ways of generating revenue from transportation to fund transport infrastructure, including maintenance of existing roads.

One option for additional taxation is a "windfall tax" on the profits of local synthetic fuel producers Sasol and PetroSA. These companies produce synthetic fuels from non-oil indigenous feedstocks (coal and natural gas) but benefit by way of increased revenues and profits when international crude oil prices rise since retail fuel prices are based on import parity pricing. A task team appointed by the National Treasury in 2006 recommended the imposition of a windfall tax on synthetic fuel producers (Rustomjee et al. 2007), as did a report compiled for the Fiscal and Financial Commission (Fofana et al. 2008). However, the Finance Minister at the time opted not to follow this recommendation, out of fear that it would undermine further investment in the synfuel industry, which he saw as necessary for bolstering energy security. As discussed in Chap. 2, however, a new CTL plant does not appear to be on the cards in any event. In essence, a windfall tax would transfer increasing resource rents from a (partially foreign-owned) private company to the country as a whole.

5.4.2 Monetary Policy

The South African Reserve Bank is mandated to maintain price stability in support of economic growth and development. The primary policy framework, which is determined by the government, is inflation targeting. The major implementation tool is the repurchase rate of interest charged on overnight loans to commercial banks, which determines short-term market interest rates. Inflation targeting goes hand in hand with a free-floating exchange rate. As of 2013, the inflation target, which is determined by the National Treasury, was set at a range between 3 and 6 % for headline inflation (annual percentage change in the consumer price index). The

NGP calls for "somewhat looser monetary policy" which is intended to "support a competitive exchange rate while continuing to target low and stable inflation" (EDD 2010). The NGP recognises that these goals are to some extent conflicting and thus identifies the need for additional policy tools to combat inflation, such as effective competition policy and probes into anticompetitive pricing and restraints on administered price increases (EDD 2010).

Peak oil presents a significant challenge to monetary policy, since oil price increases generally lead to higher rates of producer and consumer price inflation for many goods and services, and at the same time they tend to reduce economic growth, resulting in "stagflation". Critically, the SARB must recognise that (1) the inflationary pressures are largely cost-push in nature and (2) that these cost push forces will likely continue for many years. Since consumer demand will come under increasing pressure from rising costs of basic commodities (electricity, fuel, transport, and food), the SARB should avoid excessively high interest rates that destroy demand and dampen investment in mitigating alternatives. It is probably advisable to allow relative prices to change through moderately higher inflation, rather than risk serious deflation (Douthwaite 2010; Feasta 2005). As far as the exchange rate is concerned, the SARB should refrain from attempting to intervene directly in the foreign exchange markets to influence the rand exchange rate, as historical experience (e.g. in 1996 and 1998) has shown this to be ineffective and costly. Another option would be to introduce a financial transactions tax (i.e. a small percentage levy on international financial transactions), which could help to stabilise portfolio inflows on the financial account and raise much-needed revenues at the same time.

The policy measures discussed above assume that the international and national financial and monetary architecture remain intact. However, as briefly alluded to in Chap. 1, a possible scenario is that the current financial system experiences systemic collapse to one degree or another. If such an event were to occur, and arguably even if it did not, a national mitigation strategy could involve "monetary reform", i.e. a restructuring of the national monetary system. This could take the form of government-created money at very low interest rates, either by way of state-owned banks lending money into existence (as in the case of the Bank of North Dakota in the U.S.) or through the SARB being nationalised and performing the same function. The advantage of monetary reform along these lines is that investments in reducing oil dependency (e.g. in renewable energy and mass public transport) could be financed at low or negligible interest rates, along the lines of the "Green New Deal" (Barbier 2009). Critically, the long-term success of this approach would depend on a responsible state that does not create too much money relative to the goods and services available; otherwise, it could follow the path to much higher rates of inflation or even hyperinflation (e.g. as experienced in Zimbabwe in the 2000s). Once regarded as anathema to the dominant ideological paradigm and vested financial interests in market economies, similar policies have been adopted by several major industrialised economies since the 2008 financial crisis under the guise of "quantitative easing". An alternative approach to monetary reform at the local level will be discussed in Chap. 6.

5.4.3   Industrial Policy

South Africa's industrial policy as of 2013 was summarised in the Industrial Policy Action Plan 2011/12–13/4 (IPAP2), introduced by the Department of Trade and Industry (DTI) in February 2011 (DTI 2011). The IPAP2 elaborates on and supports the sectoral strategies delineated in the New Growth Path but, like that document, pays no attention to the possibility of oil supply constraints or price shocks. IPAP2 (DTI 2011: 79) prioritises three clusters of sectors, namely, (1) "qualitatively new areas of focus" (metal fabrication; capital and transport equipment; oil and gas; green industries; agro-processing; and boat-building); (2) "scaled-up and broadened interventions in existing IPAP sectors" (automotive; petrochemicals; clothing and textiles; biofuels; forestry; paper and pulp; tourism; and business process outsourcing (BPO)); and (3) "sectors with potential for long-term advanced capabilities" (nuclear; advanced materials; and aerospace). The Cluster 1 sectors have mixed compatibility with the goal of mitigating peak oil. Metal fabrication, capital equipment, agro-processing, and boat-building all are fairly energy-intensive industries, which rely to some degree on oil. The oil and gas industry in South Africa might benefit from servicing the booming West and East African oil and gas developments following recent discoveries, but this will not help South Africa to cope directly with higher oil prices unless there are new domestic fossil fuel discoveries. The manufacture of new transport equipment could help reduce oil dependency if it is focused on railways and buses, but the net energy equation needs to be carefully analysed. The "green industries" included in the IPAP2 are solar water heaters, wind turbines, solar PV panels, concentrated solar thermal, biomass energy, energy- and water-efficient materials, appliances and motors, waste management, and energy-efficient vehicles. These are mostly in line with peak oil mitigation, although the EROI of solar energy needs further investigation. In addition, these industries need to be supplemented with more sustainable transport infrastructure and vehicles, including bicycles, electric bicycles, and electric scooters. In Cluster 2, the prospects for export-oriented sectors such as automotive and tourism are dim in a post-peak oil world, while sectors with potential for import substitution such as petrochemicals (derived from coal and gas feedstock) and clothing have better prospects. The potential of biofuels again depends on the net energy they deliver, which is likely to be marginal based on international findings. In terms of Cluster 3, the nuclear value chain is highly contentious (e.g. due to cost and environmental concerns), while aerospace will likely be an unaffordable luxury. Advanced materials could become a niche area for innovation.

There are several broad policy tools that can be used to stimulate green industries. First, regulations can be imposed, such as building regulations mandating solar water heaters and energy efficiency standards for vehicles and appliances. Second, policies to support innovation are critical to the success of the mitigation programme. The NGP targets a doubling of South Africa's research and development (R&D) investment to 2 % of GDP by 2018, although the framework does not spell out how this is to be achieved. Crucially, R&D should be explicitly aimed at increasing energy efficiency. R&D can be promoted by subsidies (justified by the social benefits outweighing the private benefits) and other fiscal incentives. Key areas for technological innovation include renewable energy technologies, energy storage technologies, and more efficient vehicle designs. Third, IPAP2 correctly identifies the need for procurement policies to stimulate local production capacities (DTI 2011: 47). However, its list of "strategic procurement fleets" excludes bicycles, fuel-efficient vehicles (for government use), and solar PV panels.

The IPAP2 contends that a "continual ramp up of renewables capacity at the ambitious level of around 1–3 GW/year, would build up towards the generally acknowledged potential of at least 15 % of the electricity grid by 2020–2025" (DTI 2011: 110). However, investments in renewables need to be based on thorough life cycle and systems analyses or else they may turn out to be exceedingly costly and/ or deliver little net energy, as in the case of Spain's drive for solar power (Prieto and Hall 2013) or the US corn-based ethanol (Lambert et al. 2012). Green industries are well known to have a relatively high employment creation potential and will thus contribute to multiple socioeconomic goals (DTI 2011). However, the DTI (2011: 111) states that "[n]ational funding sources alone are insufficient to achieve a critical mass of renewable investment" and concludes that "international cooperation is required to secure the necessary concessionary finance and risk guarantee instruments".

There are several probable constraints on the implementation of industrial mitigation strategies. First, the historical trend towards increasing financialisation of the South African economy, which has diverted capital away from productive investments, will need to change (Mohamed 2010). Second, the so-called minerals–energy complex (MEC) dominates the economy and channels much of the available capital into resource-, energy-, and capital-intensive sectors (Fine 2010; Mohamed 2010). Third, the high degree of economic concentration in the economy results in monopolistic or oligopolistic pricing for many intermediate and final products (DTI 2011: 22). Fourth, financial capital for businesses to retool will likely become increasingly scarce and expensive in the era of declining world oil production. Fifth, green industries will face input bottlenecks, for example, in raw materials such as rare earth metals that are used in renewable energy technologies. Sixth, there is a chronic shortage of high-skilled workers (EDD 2010). Where possible, these constraints will need to be addressed directly as part of the mitigation strategy, for example through the application of skill development programmes. The NGP correctly identifies the need for new types of education and training to support newly emerging industries, such as green technologies and knowledge-intensive sectors.

5.4.4 Trade Policy

Since 1994, South Africa's economy has become increasingly integrated with the global economy, as evidenced by increasing shares of both exports and imports in GDP. The main thrust in trade policy, at least until recently, has been towards export-led growth. Given the future context of rising oil prices, however, trade policy authorities should anticipate the dampening effect that rising transport costs will have on both exports and imports (Rubin 2009). This suggests that they should plan for and facilitate import substitution and reduced reliance on exports, which is counter to the prevailing wisdom. In particular, subsidies to the automotive sector represent a misallocation of resources, considering that the global demand for new (ICE) vehicles is likely to contract rapidly once oil supplies begin to shrink, just as demand for cars fell substantially in South Africa and many other countries in the wake of the 2008 oil price spike and ensuing recession. Such a reorientation of trade policy will have to overcome resistance both ideologically and from vested interests that stand to lose in the short term. It is therefore essential that the scientific basis of oil depletion be properly communicated so that businesses understand the probable context for future trade.

5.4.5 Labour Market Policy

Labour market policy in South Africa has typically been highly contested by different ideological persuasions, especially on the issue of "flexibility" (i.e. the ease with which companies can hire and fire workers and the extent to which wages can adjust upwards or downwards) versus the protection of workers' rights and decent working conditions. Four areas of labour market policy should be included in the oil mitigation strategy. First, while recognising the need to protect the interests of workers, it might be advisable for government to allow greater flexibility in hiring and firing to facilitate the shift of labour out of declining sectors and into growth sectors as the economy restructures in response to rising transport costs. Otherwise, mass retrenchments might be the norm as companies are liquidated as a result of declining global and local demand. Second, given inflationary pressures from oil prices there is a strong argument for public and private sector wage restraint which should ideally occur within a social compact involving business, labour, and government. Third, education and training programmes should be aligned with the likely impacts of oil depletion, such as the decline of certain industries and the rise of others. The NGP's emphasis on artisanal and technical skills is very welcome. In addition to renewable energy, another sector that will see rapid growth is bicycle manufacture and repair. More generally, repair and maintenance skills will experience a rapid growth in demand as the "salvage economy" unfolds (Greer 2009). Fourth, employment creation could be stimulated by employment subsidies in green sectors like renewable energy and the manufacture of sustainable electric transport infrastructure. However, ever-tightening budget constraints might render subsidies unaffordable in the medium to long term. Given the long battle to improve workers' rights, some of these suggested policy changes might be fiercely resisted by trade unions if they are not fully apprised of the reasoning. Essentially, the sustainability of formal sector jobs is the critical issue.

5.5   Conclusion

South Africa's integration into the global economy since its democratic transition in 1994 has yielded tremendous benefits but has also increased its exposure to external shocks. Historically, crude oil price shocks have generally resulted in a rise in the oil import bill; a boost for some export commodities such as coal and gold, at least initially, but a decline in other exports as world demand contracts; flight of short-term capital; a depreciation of the rand exchange rate; higher rates of producer and consumer price inflation; slower (or negative) growth in real GDP; falling employment and real wages; and greater poverty and inequality. It is therefore vital that macroeconomic policies take account of the implications of the peak and decline in world oil production for the global economy and for oil price volatility. Fundamentally, there is a need to decouple the economy from oil dependence through an economic restructuring towards less oil-intensive activities and a re-localisation of production and consumption (to the extent that this reduces oil usage).

The government has a range of policy instruments at its disposal, which could in principle be modified to mitigate the impact of oil shocks. Fiscal policy is an important tool for shifting economic incentives and government expenditure patterns to promote energy efficiency, sustainability, and resilience. A windfall tax on domestically produced synthetic fuels would allow resource rents to be spent on more sustainable energy and transport infrastructure. The Reserve Bank should perhaps tolerate a moderate increase in the inflation rate as a way of enabling relative prices to adjust less painfully, for debts to be reduced, and to prevent high interest rates from causing a collapse in property values and a rise in business liquidations. Ultimately, the monetary system arguably needs to be reformed as the debt-based, interest-bearing system is inherently unsustainable. Trade and industrial policy should assume a gradual unwinding of globalisation (as applied to trade in physical goods), prepare for disruptions to global value chains, and support sustainable local industries. Labour market policy should facilitate a change in the workforce away from energy-intensive industries towards greener industries that are less dependent on oil.

Most if not all of the policy changes suggested above will bring gains for some and losses for others and are therefore likely to encounter significant opposition from groups who favour the status quo, even if the status quo is threatened by oil shocks. The following chapter addresses some of the social impacts of peak oil and suggests ways in which the most vulnerable segments of society may be protected to some degree from these.

Chapter 6
Society

A society's resilience to the economic impacts of peak oil will depend to a large degree on its existing levels of welfare, distribution of wealth, food security status, spatial configuration, and social cohesiveness. This chapter uses these aspects of contemporary South African society as a lens to understand the potential social vulnerabilities to oil price and supply shocks resulting from increasing global oil scarcity. The first section presents the status quo, while the second section considers the possible social impacts of peak oil—once again assuming a business-as-usual policy environment. The third section discusses a range of mitigation strategies and policies that could boost the resilience of South African society to the consequences of peak oil.

6.1 Social Vulnerabilities

6.1.1 Poverty and Inequality

Poverty renders people more vulnerable to economic shocks, including rising transport and food costs. Furthermore, a society characterised by a high degree of inequality can be expected to experience greater social stresses and tensions in times of economic adversity. According to the World Bank's database, 31 % of South Africa's population lived on less than $2 per day (at purchasing power parity) and 14 % were characterised as extremely poor (surviving on less than $1.25/day) as of 2010 (World Bank 2013). Although South Africa has no official income poverty line, a study undertaken for the OECD suggested an individual-level poverty line of R515 per month (Leibbrandt et al. 2010). Using this benchmark, the national *poverty headcount rate,* namely, the proportion of individuals with an income less than the poverty line was estimated at 54 % in 2008. As shown in Table 6.1, poverty rates varied greatly by population group and somewhat by gender; the poverty rate was highest for African (black) females (68 %) and lowest for white males (3 %).

Table 6.1 Headcount
poverty rate by population
group and gender, 2008

Population group	Poverty rate at R515/month (%)
African female	68
African male	60
Coloured female	36
Coloured male	35
Indian/Asian female	11
Indian/Asian male	19
White female	4
White male	3
Total population	54

Source: Leibbrandt et al. (2010)

In addition, the poverty rate was considerably higher amongst those living in rural areas (77 %) compared to urban residents (39 %).

Social grants, in the form of the Child Support Grant, the Old-Age Grant, and the Disability Grant, reach over 15 million recipients and have helped considerably to reduce the extent of poverty from earlier years (National Planning Commission 2012). Nevertheless, household income and expenditure are very unequally distributed in South Africa. The country's Gini coefficient—a measure of income inequality ranging between 0 (complete equality) and 1 (complete inequality)—stood at 0.70 in 2008 (Leibbrandt et al. 2010), which is one of the highest national rates of income inequality in the world. Inequality has been on a rising trend since 1993 (just before the end of apartheid), driven mainly by increasing inequality within the African population group as a new African middle class and elite have emerged whilst many have remained stranded in unemployment and poverty.

Poorer households are in general more vulnerable to increases in energy, transport, and food prices because a higher proportion of their incomes are devoted to these categories of expenditure. Table 6.2 shows annual expenditures on household energy, transport, and food in rands and as a percentage of total expenditure for the ten income deciles and by type of settlement in 2010/2011 (the latest available data). Food is the largest expenditure item for poorer households, although energy and transport costs are also significant. The table also shows that those households living in urban informal settlements—and to a slighter lesser extent in traditional areas—are particularly at risk of rising energy, transport, and food prices as the proportion of total expenditure spent on these items is higher than for those living in other types of settlements.

6.1.2 Household Food Security

Food security is a vital condition for human well-being and social stability. The two most important dimensions of food security at the household level are the physical availability and the economic affordability of sufficiently nutritious foodstuffs. The incidence of food insecurity in South Africa is not known with any degree of certainty owing to a lack of nationally representative surveys of that issue (Hendriks 2005).

Table 6.2 Annual household expenditure on energy, transport, and food, 2010/2011

Income	Energy		Transport		Food		Total
Group	Rands	%	Rands	%	Rands	%	Rands
Decile 1	1,012	4.5	2,279	10.2	6,843	30.7	22,300
Decile 2	1,077	4.2	2,594	10.1	8,256	32.1	25,765
Decile 3	1,347	4.3	3,203	10.3	9,314	30.0	31,103
Decile 4	1,393	3.9	3,848	10.8	9,938	27.9	35,618
Decile 5	1,583	3.8	4,933	11.9	10,443	25.1	41,607
Decile 6	1,815	3.6	5,931	11.7	11,496	22.7	50,747
Decile 7	2,143	3.1	8,482	12.2	12,213	17.6	69,316
Decile 8	2,993	2.9	12,177	11.6	14,519	13.9	104,788
Decile 9	4,513	2.6	17,125	9.8	16,881	9.7	174,281
Decile 10	7,062	1.8	27,847	7.0	22,091	5.6	396,095
Total	2,494	2.6	8,843	9.2	12,200	12.8	95,183
Urban formal	3,081	2.3	11,581	8.8	13,296	10.1	131,297
Urban informal	1,528	4.4	4,949	14.4	8,472	24.6	34,481
Traditional area	1,543	3.7	4,681	11.4	10,955	26.6	41,183
Rural formal	2,778	3.2	7,142	8.3	13,182	15.4	85,559

Source: StatsSA (2012)
Notes: Energy includes electricity, gas, liquid fuels, and solid fuels used in homes; transport includes operation costs (including fuel) plus transport services; food includes foodstuffs plus non-alcoholic beverages

Nevertheless, the available evidence indicates that in the vicinity of 59–73 % of households experience food insecurity, approximately 16 % have an inadequate energy intake, about 30 % experience hunger, and approximately 22 % of the population could be stunted and 3.7 % afflicted by wasting (Hendriks 2005: 115). While the determinants of food insecurity are numerous and complex, two basic drivers can be identified, reflecting the principal dimensions referred to above. These are, firstly, inadequate income (i.e. poverty) in relation to the cost of food products and, secondly, a lack of access to land, water, and other productive inputs required for own food production. Somewhat ironically, however, hunger and malnutrition in South Africa are more common in agricultural areas, especially amongst smallholder farmers, agricultural workers, and the landless (Hendriks and Msaki 2009: 185). This is essentially because the incidence of poverty is higher in rural areas (Pauw 2007). Oil price shocks carry two major threats to household food security. The first is rising food prices, since food prices are linked to oil prices through input and transport costs. The second threat is of falling employment levels and incomes as a result of the negative macroeconomic consequences of oil shocks. These factors act together to reduce the food purchasing power of households.

6.1.3 Settlement Patterns

Human settlement patterns play an important role in South African society's dependence on liquid fuel-based transport and therefore in its vulnerability to peak oil.

The two predominant features of the country's settlement patterns are inequality and inefficiency, and the two aspects are closely linked for historical reasons. Inefficiency in the spatial configuration of settlements relates largely to the phenomenon of urban sprawl, which occurs in virtually all cities and metropolitan areas around the country. Approximately 60 % of South Africans live in urban areas (Leibbrandt et al. 2010). Two forms of urban sprawl characterise South Africa's cities. First, as a result of apartheid-based urban settlement patterns characterised by remotely located townships, many lower income citizens tend to reside in informal or formal settlements that are located on urban peripheries. This pattern was perpetuated by an ongoing process of urbanisation, which accelerated after the democratic transition in 1994. The key problem is that these settlements are located far from social and economic opportunities (The Presidency 2008: 99). The large distances impose high travel costs, which are particularly burdensome for residents who are generally poor to begin with. Second, the majority of more affluent citizens reside in sprawling suburbs that are highly dependent on private motor vehicles. Both groups generally need to travel substantial distances to access economic and social opportunities and are therefore vulnerable to rising fuel and transport costs and fuel shortages. Nevertheless, rural areas are in some ways even more dependent on motorised travel than urban areas, as distances to towns may be larger than distances typically travelled within cities, and there are less extensive or even nonexistent public transport services.

6.1.4　Social Cohesion

The ability of a society to withstand social and economic shocks depends *inter alia* on the degree of social cohesion. Social cohesion in South Africa is affected by numerous factors including the ethnic diversity of the society, actual or perceived racism, political dispensation, inequality and poverty, migration and immigration, and rates of crime and violence. South Africa has enjoyed a stable democracy since 1994 with free and fair elections held every 5 years since then. The African National Congress (ANC) has maintained power throughout this period with a strong majority in the National Assembly. There has been no widespread, serious political violence in the democratic era. Nevertheless, social tensions persist after nearly 20 years of democracy, reflecting in part the legacy of a conflict-ridden apartheid past and also the persistence of income inequality and poverty (The Presidency 2008: 29). Domestic economic and social stresses have been further compounded in recent years by an increased rate of immigration from other African countries (The Presidency 2008: 99). Immigration has been driven both by push factors (such as political, social, and economic upheavals in their home countries) as well as pull factors (e.g. the relative size and strength of South Africa's economy in the context of Southern Africa). This immigration has placed extra strain on already overstretched social services and aggravated social tensions, as evidenced by the socalled xenophobic violence that erupted in many parts of the country in May 2008

(Sharp 2008). More generally, South African society is characterised by unacceptably high rates of crime, especially violent crimes such as murder, rape, assault, and robbery (The Presidency 2009: 59). The causes of crime are complex and multifarious and include the high rate of unemployment, especially among young males, as well as the related high incidence of inequality and poverty.

6.2   Possible Social Impacts of Peak Oil

A progressive, sustained contraction in the economy, together with growing restrictions on mobility, would place great strain on South African society. Changes in household poverty and inequality will largely be determined by the macroeconomic impacts of declining world oil production, which are likely to include economic recession, higher price inflation (especially for food and transport), rising unemployment, and falling real incomes. These trends will all contribute to a rise in the incidence and depth of poverty. Those who are already poor will suffer the most as they spend a high proportion of their meagre incomes on food and fuel (including paraffin for cooking) and lack the requisite resources to adapt. Those who lose their jobs are at the greatest risk of joining the ranks of the poor. On the other hand, rising costs of petrol and diesel have the greatest direct impact on middle-income households, since poorer households cannot afford private vehicles. Pensioners who rely on fixed incomes and equity investments will find their real incomes eroded by the rising cost of living and falling value of equities. Tighter government budgets may constrain government from increasing the value of social grants to match the rate of price inflation, resulting in falling real incomes for the approximately 16 million citizens who depend on social grants. Municipalities are also likely to find their budgets under increasing pressure, both from the revenue side (if transfers from national government fall and defaults on payments of rates rise) as well as the cost side (e.g. rising expenditures on transport and road maintenance). Tighter municipal budgets would tend to negatively affect service delivery, compounding socioeconomic hardships. Empirical models suggest that income inequality is also expected to rise as a consequence of repeated oil shocks (Essama-Nssah et al. 2007).

In the medium term, household food security will deteriorate as rising food prices combine with falling real incomes to reduce the affordability of food. This will lead to increasing rates of hunger and malnutrition and growing demands on state food support systems. In the longer term, the physical access to food could also be compromised as a result of declining domestic agricultural production (see Chap. 4) and because fuel shortages and immobility will likely disrupt the distribution of food products to consumers in urban areas.

The direct consequences of increasing oil scarcity, such as rising transport costs and restrictions on mobility, as well as the indirect impacts via changing economic conditions, will have significant consequences for patterns of human settlement and migration. In the short term, most people will respond to rising transport costs by adapting as best they can where they currently live. In the medium term, higher fuel prices will incentivise people who have the means to move closer to employment opportunities, schools, etc. The rate of rural–urban migration might increase as economic conditions are likely to deteriorate faster in peripheral rural areas than in urban areas. In the longer term, however, this migration pattern may shift into reverse, with a decanting of urban populations into rural areas (Heinberg 2006b). This would likely be driven by a rising incidence in crime and violence in crowded cities where social services are deteriorating and also by food shortages forcing people to return to the land to grow their own food, provided they could afford to buy property or gain access to communal land.

In the scenario of progressive economic contraction, satisfaction of poor people's basic needs will increasingly be in jeopardy as joblessness and hunger increase, while the provision of social services will be hampered by dwindling tax revenues and mounting costs. This combination of factors is likely to spark further increases in social protests, as experienced since 2008. The high levels of poverty and inequality, high levels of violent crime, and recent episodes of xenophobic violence raise the possibility of large-scale class or ethnic conflict, especially in cities, and possibly a localised breakdown in the rule of law. This social fragmentation could be exacerbated by immigration from neighbouring countries, whose people may be attracted to South Africa's relatively stronger economy. During the 2000s, for example, South Africa received a flood of refugees from neighbouring countries suffering from the effects of droughts, HIV/AIDS, and especially Zimbabwe's economic collapse.

6.2.1 Systemic Impacts

It is important to recognise the systemic vulnerabilities of our complex society to external shocks. The national socioeconomic system comprises several interconnected subsystems, such as energy, transport, communication, financial, water, sewage, and food systems, each relying on interdependent critical infrastructure. Like other complex systems, they are characterised by thresholds, tipping points, and positive feedback loops, so that shocks to one subsystem can have non-linear impacts—both within that subsystem and cascading through other connected subsystems. The possibility and mechanisms for a systemic collapse of complex societies have been examined by several notable scholars (Diamond 2005; Tainter 1988) and usefully explored in relation to peak oil by David Korowicz (Korowicz 2010a, b). Applying Korowicz's logic to the South African socioeconomic system highlights the possibility of systemic crises. Figure 6.1 represents schematically the linkages between six of the major critical systems operating in the South African economy as well as their connections to human welfare and social cohesion. Peak oil will impact most directly on the transport and financial systems (the latter via the global and local economies). The transport system affects electricity generation (e.g. via truck deliveries of coal to power stations), supply chains (distribution of components and final products), food (production, processing, and distribution to

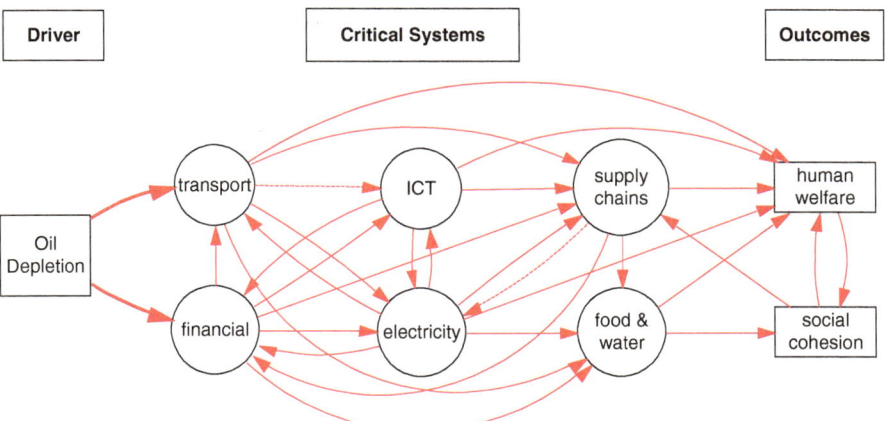

Fig. 6.1 Flow diagram of possible or probable causative chains in critical system linkages. *Key*: *bold arrow* primary causative force, *solid arrow* short-term impact transfer, *dotted arrow* medium/ longer term impact transfer

markets), and in the longer term information and communication technology (ICT) by making maintenance of infrastructure more costly. The financial system affects all five other critical systems by facilitating economic investments and transactions. The electricity system enables the functioning of all the other systems, either wholly (e.g. ICT) or partially (e.g. food via irrigation, processing, and cold storage). The ICT system is integral to the functioning of the financial, electrical, and supply chain systems. Economic supply chains are critical to the food system and link back to the health of the financial system and (in the medium term) to the electricity system (by supplying components required for maintenance). Human welfare is dependent on the transport, ICT, supply chain, and food systems, while the major determinants of social cohesion are human welfare and access to food. The state of social cohesion feeds back to human welfare and the functioning of supply chains. A major crisis in any one of the interlinked subsystems will be transmitted to other subsystems until it reverberates through the entire socio-economy. In a worst-case scenario, such as a widespread failure of the electricity grid or a systemic freeze of the banking sector, the resulting hunger, rioting, looting, and a breakdown of the rule of law could lead to a societal fragmentation within just a few days (Simms 2008).

6.3   Boosting Societal Resilience and Cohesion

In the social domain, the main goals for a peak oil mitigation strategy must be to protect the most vulnerable members of society and to promote equity. More specific objectives include promoting fairness in access to fuels; building community resilience to socioeconomic shocks; enhancing household food security; strengthening social cohesion; and planning to ensure sustainable spatial development in the long term and an ability to cope with emergency situations brought about by fuel supply disruptions. Ways in which these objectives can be met are discussed in the following sections.

6.3.1 Social Protection Systems

Allowing market prices to determine who in society gains access to fuels brings no guarantee of fairness. The wealthy will still be able to afford fuel for their private automobiles, while poorer people may not be able to afford higher public transit fares. Unfortunately in recent years there has been an increase in the Gini index for South Africa, indicating that income inequality is increasing, probably partly in response to the increasing global emphasis on "the market" as a "virtuous and fair" means of resolving distributional issues. This inequality could precipitate serious social strife. To ensure fairness in access to and/or affordability of fuels and energy services, some form of administrative system should therefore be implemented. The pros and cons of four options are considered below, namely, fuel price controls, liquid fuel coupons, tradable energy quotas (TEQs), and a "cap and share" system.

6.3.1.1 Fuel Price Controls

Government can in principle impose domestic fuel price controls to limit the rise in fuel prices to international levels (Hirsch et al. 2010: 87). In effect this means imposing price ceilings on domestic oil producers Sasol and PetroSA. Since approximately two-thirds of petroleum products are refined from imported crude oil, the government would also have to introduce an entitlement system for oil refiners that averaged the prices of domestic and imported oil. Thus refined fuel prices would still rise, albeit not to the world level. This system requires additional price controls on the markups of downstream operators such as refiners, wholesalers, and retailers; in South Africa these are already part of the existing regulatory regime. In the absence of allocation controls (rationing), price controls would result in long queues at filling stations since prices would not be high enough to reduce demand to the lower level of supply. On the plus side, all fuel consumers would benefit from the attenuation of fuel price increases, while domestic producers would not capture excessive scarcity rents. However, price controls have several major disadvantages. First, queuing for fuel is highly inefficient and results in a net welfare loss to society (Hirsch et al. 2010: 91). Second, the poorest members of society would still face price increases that would render fuel or transport increasingly unaffordable. Third, price control systems are administratively onerous, requiring refiners to submit detailed information to the authorities on quantities of oil bought and sold, and a large team of government personnel would be needed to administer the system. Fourth, such a system lacks the flexibility to cope efficiently with fluctuating oil prices. Finally, fuel price controls could potentially undermine the economic viability of the domestic producers, if their operational costs rise as a consequence of rising world oil prices.

6.3.1.2 Liquid Fuel Coupons

In order to overcome the queuing problems discussed above, price controls can be accompanied by a rationing system. Rationing systems have been adopted in many countries in the past, both for oil and for other essential commodities. The most basic rationing system involves booklets of coupons issued to citizens on the basis of registered vehicle ownership, historical consumption patterns, and/or priority users (Hirsch et al. 2010: 88). Citizens would have to relinquish their coupons (e.g. denominated in litre units) every time they purchased petrol or diesel; retailers would pass on coupons to refiners; and refiners would transfer the coupons to the government. Such a rationing system would have to be complemented with domestic oil price and refined fuel price controls as above or else petrol and diesel prices would rise to market-clearing levels. A secondary market for fuel coupons would develop, so that citizens requiring (and being able to afford) more fuel could purchase ration coupons from those needing less. The main advantages of a rationing system are its fairness, the elimination of queuing, and the opportunity for the government to allocate coupons to priority users such as emergency services and those with particular hardships or high fuel dependency. However, such simple systems have many drawbacks, including high administrative costs and human capacity requirements; interest groups exerting pressure on government officials responsible for coupon allocation; possible forgery of coupons; greater complexity if companies and state institutions are included; economic inefficiencies associated with price controls and allocation; and a limitation of ration trading opportunities to local areas (Fleming 2007; Hirsch et al. 2010).

6.3.1.3 Tradable Energy Quotas

The concept of TEQs was developed specifically to address the twin challenges of climate change and fossil fuel depletion while ensuring social equity in access to energy (Fleming 2007). TEQs function as follows. To begin with, a Climate Committee determines a national "Carbon Budget" for total annual carbon dioxide emissions, which decrease each year over a 20-year period. The annual budget is then divided into individual TEQ units, a unit being "defined as one 'carbon unit'— that is, allowing the purchase of sufficient fuel or energy to produce one kilogram of carbon dioxide over its lifecycle" (APPGOPO 2009: 9). TEQs are required for all energy purchases, with each energy type being rated according to its carbon content. A registrar issues equal entitlements (summing to approximately 40 % of units) to each adult citizen and maintains carbon accounts for all participants. In addition, TEQs are sold on a weekly tender via financial intermediaries to other energy users, including businesses and government. Once issued, TEQs can be traded by all energy users in a single market, allowing low-energy consumers to derive income while high-energy users purchase sufficient units to cover their energy needs. The entire system is electronic, making use of direct debits and electronic cards. The price of TEQs is determined in this market and depends on efforts to reduce energy demand. It is not necessary to measure emissions at their exit point ("exhaust pipe"); rather, the carbon accounting is done at the point of sale. When individuals or entities purchase energy, they surrender the appropriate number of units to the retailer, who surrenders to wholesalers, etc. Primary energy producers surrender units to the registrar. Each successive year, the total Carbon Budget is reduced and thus fewer TEQs are issued.

TEQs have a number of benefits. First, they guarantee a certain annual reduction in fossil fuel use and carbon emissions. Second, TEQs promote fairness in access to energy while providing strong incentives for individuals to reduce their energy consumption. Third, a rolling carbon budget is set for a 20-year period, which creates a much greater degree of certainty for business and personal planning decisions. Fourth, the system is administratively simple, using modern, automated electronic systems that make it largely "hands free" (Fleming 2007). Nonetheless, TEQs do have certain limitations in the South African context. For one, TEQs require a single national market for carbon-based fuels, but this is not necessarily simple; for example, the incorporation of electricity generated from fossil fuels is complicated when renewable primary energy sources are also used. Second, TEQs do not directly protect individuals against high oil prices, which are still determined on international markets. Thus there is no guarantee that poorer people will actually be able to afford the fuel that their TEQs entitle them to. Nonetheless, they will be guaranteed a basic income through their right to sell their TEQs. Third, implementation of the TEQ system would face challenges in a developing country context such as South Africa, where 40 % of the population did not participate in the formal banking system (African Loft 2008) and 23.5 % of adults were illiterate in 2008 (The Presidency 2010: 48).

6.3.1.4 Cap and Share

A similar scheme, known as Cap and Share (C&S), has been developed by the Ireland-based Foundation for the Economics of Sustainability (Cap and Share 2011; Feasta 2008). Like TEQs, C&S was designed to address the related challenges of peak oil and climate change. It can in principle be applied on a global or a national scale. Under a national C&S scheme, an independent scientific body determines annually a maximum ceiling (cap) on carbon dioxide emissions from the combustion of fossil fuels. The total amount of emissions is divided equally among the adult population; each citizen is allocated a "pollution authorisation permit" (PAP), i.e. a certificate entitling the owner to emit a certain tonnage of CO_2 per year. Individuals are then free to sell their PAPs through financial institutions (e.g. banks and the post office) in a national market for PAPs. Fossil fuel producers and suppliers (e.g. coal miners, oil companies, and gas importers) must buy sufficient PAPs to cover the emissions from the coal, oil, or gas they sell into the economy. These companies then surrender these PAPs to an independent agency that monitors the actual CO_2 content of the fuels. Each year the cap is reduced by a percentage determined by the scientific panel (or faster than the rate at which global oil production declines). The fossil fuel suppliers will pass on the additional cost of the PAPs to consumers (ultimately to consumers of final energy products like petrol and electricity), but consumers will be at least partially compensated by the income they derive from selling their PAPs. Individuals who consume less than average fossil fuel-based energy might benefit in net terms, while those consuming more energy will pay more in net terms.

In a variant of C&S, known as Cap and Dividend (C&D), an independent agency conducts auctions to sell the PAPs to fossil fuel suppliers and distributes the monetary proceeds on an equal per adult basis. This variant would be better suited to the South African context of relatively high rates of illiteracy and lack of education, as individuals would not have to make complicated decisions about when to sell their PAPs. A portion of the auction dividends could be earmarked for specific community or social investments, such as renewable energy production or public transport systems.

C&S is designed to transfer the scarcity rents on fossil fuels from producers to all adult citizens in an equitable manner. It also incentivises people to reduce their consumption of fossil fuels both directly (as fuels) and indirectly (embodied in goods) and simultaneously boosts the competitiveness of renewable energy sources. The diminishing annual cap means it would be effective in reducing CO_2 emissions. Since there are few fossil fuel companies bringing their products into the economy, the system would be simple to administer. Monthly dividends could be distributed in a manner similar to existing social grants, i.e. through post offices. In fact, these dividends would effectively constitute a "basic income guarantee" for all adult citizens and thereby provide a measure of social protection against the negative socioeconomic impacts of global oil depletion. On the other hand, C&S would not function as desired if there were rapid declines in oil imports, and therefore other emergency rationing schemes would be required under such circumstances.

The major drawbacks to all rationing systems are the administrative costs and human capacity required and the risk of corruption. Of the schemes discussed above, Cap and Dividend is the simplest and cheapest. TEQs, being a downstream system, are more complicated and expensive to administer than C&S or C&D, which are upstream systems (Cap and Share 2011). TEQs and C&S (or C&D) have other advantages over fuel rationing, in that they serve a wider range of purposes (e.g. climate mitigation). They are also more flexible, for example by including non-fuel energy such as electricity derived from coal, which is also likely to become increasingly expensive as coal reserves are depleted. However, fuel rationing has the advantage that specific quotas can be distributed to priority users such as essential services (e.g. police, fire brigades, and other emergency services), public transport vehicles such as buses and diesel trains, and farmers. It is also more effective in the event of sudden, large fuel shortages (e.g. triggered by geopolitical events in oil-producing nations that shut in large amounts of supply). At a theoretical level, schemes such as TEQs and C&S reflect the fundamental role that energy plays in sustaining economic activity, by bringing energy closer to the centre of economic transactions. Fuel rations, TEQs, or PAPs effectively become alternative currencies that could help to mobilise resources when the supply of conventional money is scarce, a topic discussed further in the following section.

6.3.2 Building Community Resilience Through Economic Localisation

Although national economic policies and conditions affect economic development throughout the country, all development actually takes place at the local level. Furthermore, peak oil will in all likelihood result in a gradual reversal of globalisation in respect of trade in physical goods and transport-dependent services such as tourism (Rubin 2009). Thus any mitigation strategy must include a strong local dimension, the specific goals of which are to (1) improve community resilience to economic shocks and facilitate sustainable development at a local level; (2) boost the number of sustainable livelihoods; and (3) enhance household food security. Local economic development (LED) strategies and policies operate at the local (municipal) level of government, in both urban and rural areas, and complement national economic policies. This section explores instruments that can be used to achieve the above goals.

The Department of Cooperative Governance and Traditional Affairs (DCoGTA 2011a) runs an LED programme guided by the National Framework for Local Economic Development (NFLED), the overarching aim of which is to "support the development of sustainable local economies through integrated government action" (DPLG 2006: 3). Within this framework, LED is understood to be "about creating a platform and environment to engage stakeholders in implementing strategies and programmes" and involves "the provision of infrastructure and quality services" by municipalities with the aim of improving competitiveness of local economies (DPLG 2006: 9). The NFLED identifies two main policy thrusts. The first is "sound public sector leadership and governance". The second is "sustainable developmental community investment", which aims to boost "the circulation of income in local economies". This is to be achieved via community trusts that will accumulate both public and private investment capital and spend these on community projects via cooperative enterprises. The NFLED also calls for increased spending on "adult basic education and training aimed at improving literacy and numeracy as a basis for participating in local economic opportunities", particularly in rural areas (DPLG 2006: 14). Local government can further assist LED through development marketing bodies and investment incentives.

For the most part, LED strategies in the past have assumed that the forces of globalisation will be maintained or even increase in the future and seek to counterbalance these forces or adapt to them. Eco-localisation (or "re-localisation") calls for a proactive strategy to reduce reliance on a globalised economy that is expected to "de-globalise" as a consequence of rising transport costs as world oil production declines (Heinberg 2004; Hopkins 2008; North 2010). Localisation implies "producing as much as possible as locally as possible, then within the shortest possible distance, with international trade only as a last resort for goods and services that really cannot be produced more locally" (North 2010: 587); intra-industry trade would largely disappear. Since global value chains are highly susceptible to dislocation from liquid fuel shortages, businesses that are currently embedded in global and even national value chains should, wherever possible, seek to localise their sources of inputs and expand their local markets. Regional and local multipliers can be boosted by measures such as "buy local" campaigns (e.g. the "Proudly South African" initiative). Further than this, intentional localisation "means developing community-owned local economic institutions like worker-owned and run cooperatives, communal gardens and restaurants, local power generation, local money, and communal forms of land ownership" (North 2010: 587). Proponents of localisation advocate local economies that are resilient through diversity and interdependence. Authors such as Hopkins (2008) foresee that re-localisation will be driven primarily by communities, although working in partnership with local government authorities. The "Transition Town" movement, inspired by Hopkins, is putting this idea in practice as communities formulate "energy descent action plans" in order to manage the forthcoming decline in energy availability. Similarly, the doctrine of "eco-localism" holds that "[t]he road to environmental sustainability lies in the creation of local, self-reliant, community economies" (Curtis 2003: 83). According to Curtis, eco-localism "is the perspective embodied in local currency systems, food co-ops, micro-enterprise, farmers' markets, permaculture, community supported agriculture (CSA) farms, car sharing schemes, barter systems, co-housing and eco-villages, mutual aid, home-based production, community corporations and banks, and localist business alliances". In addition to conventional markets, local economies are "equally constituted by collectives and cooperatives, buying clubs, community enterprises, not-for-profits, barters and skills exchanges, mutual aid, volunteer activity, household and subsistence production, and what is variously termed the informal sector or the underground economy" (Curtis 2003: 86).

The second major goal of an LED or a re-localisation strategy is to create sustainable livelihoods. The National Strategy for Sustainable Development (NSSD) identifies opportunities for sustainable livelihoods in eco-tourism, aquaculture, small-scale organic agriculture, ecosystem rehabilitation, renewable energy generation, and wildlife management (DEA 2010: 18). The NSSD further proposes that the Public Works Programme be extended into the environmental sector. To some extent this has been achieved through the Expanded Public Works Programme (EPWP), which includes four sectors, namely, infrastructure; environment (including the Working for Water, Working for Wetlands, Working on Fire, and Working for Energy programmes); non-state sector; and social sector (DPW 2011). The EPWP creates temporary work for unemployed citizens but aims to build skills that will help individuals to find sustainable work opportunities. The EPWP Phase 2, launched in April 2009, aimed to generate two million full-time equivalent jobs over the ensuing 5-year period. The EPWP has successfully created short-term work opportunities but has been less successful in creating sustainable livelihoods and is a relatively costly programme (Philip 2010).

The Department of Cooperative Governance and Traditional Affairs runs a Community Work Programme (CWP), which aims to provide an "employment safety net" to improve the quality of life of people living in marginalised areas (DCoGTA 2011b). The CWP is implemented in specific sites (in wards or municipalities) using a community participation approach to determine what constitutes "useful work", defined as "work that contributes to the public good and improves the quality of life in communities" (TIPS 2010: 5). Practical examples of useful work include home-based care, food gardens, community parks and gardens, rehabilitation of ecosystems, road maintenance, early childhood development programmes, and community safety. The first phase of the CWP was considered to be very successful in meeting its objectives and could serve as a basis for a national employment guarantee (Philip 2010), which in turn could prove to be a vital tool for mitigating the socioeconomic impacts of peak oil. As Philip (2010: 27) argues, "an employment guarantee doesn't only contribute to raising aggregate demand at local level: it also invests in human capital development, in public/community goods and services, and in natural capital, in ways that further enhance the potential for sustained social and economic development". Philip (2010: 24) estimates that an expanded CWP would cost about R10 billion per million participants, less than half as much as the EPWP. The key issue is a sustainable funding model, which is explored at the end of this section.

The third goal of boosting food security can be addressed through the expansion of small-scale sustainable agriculture, which can simultaneously assist in the creation of sustainable livelihoods. According to TIPS (2010: 6), "[t]housands of food gardens have been established through the CWP, making a huge difference to food security at a household level, and providing free food for feeding schemes and vulnerable households". Hine et al. (2008: 12) report that the adoption of organic farming techniques amongst smallholder African farmers is associated with improving food security. It has been argued that organic farming can boost food security by raising productivity, generating safer food, attenuating production costs, reducing risks through diversification, supporting innovation, and enhancing long-term sustainability (INR 2008). A critical precondition for the success of sustainable (e.g. organic) smallholder agriculture is the acquisition of appropriate knowledge and skills by emerging farmers; here the state has an important role to play. Furthermore, international evidence suggests that land redistribution can be an effective way of improving the welfare of landless rural dwellers, provided the land is of sufficient quality (Rosset 1999). The Comprehensive Rural Development Programme (CRDP), published in July 2009 by the Department of Rural Development and Land Reform (DRDLR), "is aimed at being an effective response against poverty and food insecurity by maximizing the use and management of natural resources to create vibrant, equitable and sustainable rural communities" (DRDLR 2009: 3). The CRDP strategy involves three elements, namely, agrarian transformation; rural development (stimulated by strategic investment in economic and social infrastructure); and land reform. In practice, however, the CRDP was limited to approximately 56,700 households residing in 21 wards in 2009/2010 and envisages expansion to include 121,500 households in 45 wards by 2013/2014. The land reform process in South Africa is lagging far behind targets (DRDLR 2009). The success of this process will be even more important in the future context of fuel scarcity, which as noted earlier is likely to reverse the trend towards mechanisation of farming and consolidation of farms into larger units; that is to say, more labour will likely be required on farms in the future.

A crucial element of an effective LED or re-localisation strategy concerns ways to boost the local money supply, stimulate local money multipliers, and stem monetary leakages. Funds for LED could in principle be sourced from loans from national financial institutions or grants from national government and donors, but these sources will tend to be increasingly unsustainable in the era of declining world oil production. Three local approaches to ensuring a sufficient supply of money have much better prospects. The first option is the formation of state-owned municipal banks, which would be able to award low- or zero-interest loans to individual citizens and for community investment projects, as the state-owned Bank of North Dakota does in the United States (Harkinson 2009). The second option is the formation of credit unions, which are cooperative, community-owned financial intermediaries. These institutions essentially pool the financial resources of citizens and make loans on favourable terms for projects adjudged by the governing board of directors to be sustainable and beneficial to the community (Douthwaite 1996). The third option is the use of local or community currencies, which are complementary to the national currency (Douthwaite 1996). An effective network of community currencies is already operated by the Community Exchange System, a non-profit organisation that began in Cape Town in 2003. As of September 2013, the CES network included 593 local exchanges in 60 countries, including 45 exchanges in South Africa and 170 in crisis-wracked Spain (CES 2013). Essentially, all of these mechanisms would facilitate the mobilisation of otherwise idle natural and human resources (labour and skills) on an ongoing basis to produce goods and services that are needed by the community but that might not otherwise be produced as a result of lack of access to (externally originating) money. They also help to stem leakages of money outside of the local area and hence boost local economic multipliers.

6.3.3 Spatial Development Planning for Sustainable Human Settlements

The South African Government's main spatial development planning policy document (the National Spatial Development Perspective of 2006) fails to take into account the implications of peak oil. And yet the growing scarcity of easily accessible and cheap oil calls into question many prevailing assumptions that are based on historical trends and creates significant uncertainty regarding future migration patterns and the viability of the existing stock of residential, industrial, and commercial buildings. Growing oil scarcity implies three key principles for spatial development planning. First, all new housing, commercial, and infrastructural developments should assume increasing prices and scarcity of oil-based transport fuels in the foreseeable future (e.g. a 20–30-year time horizon). Second, spatial development planning must be integrated with transport planning as well as with industrial and trade policies (see Chaps. 3 and 5). Third, planning should be based on an assumption that urbanisation will continue (and possibly accelerate) in the short to medium term but may be replaced by re-ruralisation in the longer term.

A number of specific policies and measures should be adopted in urban planning. First of all, the phenomenon of urban sprawl should be halted in order to avoid further entrenching dependency on automobile transportation and to limit encroachment on agricultural land surrounding urban settlements. Rather, urban densification "in well-located areas" and "more intense development of vacant and underused land" should be a major policy focus (Turok et al. 2011: 19). Mixed-use zoning and allocation of existing open spaces should be applied to facilitate localisation of food production. More generally, agriculture should be integrated with urban development planning, based on an ecosystem approach (Thornton 2008). Local government authorities should allow and encourage small commercial outlets, which might be termed "neighbourhood centres", within residential areas (City of Portland 2007: 37). Planning should encourage housing of sufficient density near public transport routes that will generate adequate fare revenues to sustain the mass transit services, foster developments along public transport corridors, and provide public spaces accessible to pedestrians in urban centres. Special efforts should be made in the short to medium term to maintain existing industrial areas, especially those that are near rail lines, to facilitate local economic diversification in the event that globalisation unwinds in the longer term (City of Portland 2007: 38).

6.4   Conclusion

The extent of poverty, inequality, unemployment, and food insecurity in South Africa poses enormous social challenges as it is, but these pressures will become even more intense once global oil production enters its decline phase and oil prices rise further. Without an adequate response by national and local government, society faces the risk of a cascading collapse of critical systems and social fragmentation. Fortunately, there is a broad range of policies, measures, and programmes that can be implemented to mitigate the social impacts of peak oil. Some form of fuel or energy rationing system will be essential to promote equity in access to fuels and transport and thereby foster social cohesion. Economic localisation can be promoted in order to strengthen community resilience to oil-related shocks. There are many opportunities for sustainable livelihoods to be created, for instance in small-scale organic agriculture, preservation of ecosystem services, and localised manufacturing. Finally, long-range planning for human settlements must be properly informed by the reality of global oil depletion and all it implies for our oil-based urban infrastructure. This challenge of transitioning the socioeconomic system to a regime that is less oil dependent and more sustainable is explored in more detail in the following chapter.

Chapter 7
Can We Transition to Sustainability?

The previous chapters have spelled out the wide-ranging vulnerabilities of South African society to peak oil and the likely impacts of oil price and supply shocks if our government and society continue along a business-as-usual trajectory. A strong pragmatic case has been made for the government to implement mitigation strategies to lessen the negative consequences of increasing oil scarcity for social welfare and economic development. One can further argue that the state has a clear moral and constitutional obligation to mitigate peak oil, to limit its detrimental societal impacts, and to help avoid possible economic and social collapse. It is worth noting that mitigation of peak oil in a developing country context such as South Africa is arguably less politically and morally contentious than mitigation of climate change. For developing countries the goal of mitigating climate change, specifically by reducing greenhouse gas emissions from the combustion of fossil fuels, is often regarded as inimical to the objectives of socioeconomic development and poverty alleviation (at least for the current generation). The main reason given for this viewpoint is that fossil fuels are cheaper than alternative energy sources and thereby allow more rapid and extensive industrialisation and rising living standards. In contrast, mitigation of peak oil does not require deliberately retarding economic growth or forgoing a cheaper industrialisation path, since oil will become increasingly scarce and expensive, rendering an oil-intensive industrialisation process increasingly unviable. Furthermore, many of the peak oil mitigation strategies and policies advocated in preceding chapters are expressly designed to ameliorate the negative impacts of peak oil on poverty, employment, and human well-being.

Supposing that the South African Government did recognise that peak oil represents a threat to the country's developmental aspirations, some might argue that the best response would be to capitalise on the country's substantial coal reserves and possibly large shale gas resources, together with the world-leading coal-to-liquid and gas-to-liquid technological expertise of its petrochemical companies Sasol and PetroSA. A CTL- and GTL-intensive strategy, whereby the government would give policy and financial support to expanded synthetic fuel production aimed at making South Africa self-sufficient in liquid fuels, could theoretically preserve the economic and political status quo for many years. It could be argued that this would be an efficient use of domestic resources that would give the country time and funds to develop renewable energy and electrified transport infrastructure.

However, a CTL- and GTL-intensive mitigation pathway has serious ethical and pragmatic drawbacks and risks. The ethical question that such a path raises is related to climate change mitigation, since CTL and GTL processes are highly carbon intensive. The voluminous extra greenhouse gas emissions that would be associated with expanded synfuel production could make South Africa an international pariah should the rest of the world embark on concerted action to mitigate climate change. Economically, such increased emissions could seriously harm this country's export sectors if there were international greenhouse gas regulatory regimes such as a cap and trade system or carbon taxes. Therefore this path would likely be one of increasing international isolation, echoing the apartheid era (when CTL technology was first developed). Domestically, the negative environmental and health impacts associated with coal mining, shale gas fracking, and fossil fuel combustion are also important ethical considerations, which involve differential intra- and intergenerational welfare impacts. While most of the benefits of enlarged CTL and GTL production would accrue to the current generation of South Africans (and especially to those able to afford motor vehicles), the costs would be borne disproportionately by local communities in the vicinity of hydrocarbon resources and synfuel plants, together with future generations of South Africans, who would be burdened with the long-term environmental and health consequences.

In addition to these ethical concerns, there are serious pragmatic questions about the feasibility and desirability of a synfuel-intensive path. Firstly, the recent studies cited in Chap. 2 that evaluate South Africa's remaining coal reserves and plausible production profiles indicate that the country's coal production could peak around 2020. Furthermore, demand for coal to feed new CTL plants would run into increasing competition with coal export markets and demand from power utility Eskom. At the very least, this would significantly boost coal feedstock prices and after a number of years may result in serious physical shortages of coal for one or more of these uses. Secondly, as mentioned earlier, water shortages and geological conditions in the Waterberg area could seriously inhibit the exploitation of these coal fields, which constitute the bulk of South Africa's remaining resources. Thirdly, the extent of commercially viable shale gas resources will become clear only after extensive exploration that will take perhaps a decade. Fourthly, pursuing a synfuel-intensive path would entrench the nation's lock-in to a mineral- and energy-intensive development path, with its associated political economy ramifications in terms of even greater dominance of the minerals–energy complex (MEC). It would also delay the adoption of renewable energy technologies and squander an opportunity for South Africa to develop domestic renewable energy manufacturing capacity while the rest of the world forges ahead in this growth industry. Finally, spending hundreds of billions of rands building capital-intensive plants to produce fuels for relatively inefficient internal combustion engine vehicles could be a massive misallocation of resources. Instead, the nation's remaining fossil energy reserves could be used to create a more sustainable energy and transport infrastructure—one characterised by efficient, integrated, and electrified mass transit modes powered ultimately by renewable energy.

In short, further entrenching our dependence on petroleum fuels would delay the inevitable transition to sustainable forms of energy that one day must occur. Morally and in some ways practically, it would be analogous to the conservative apartheid government's refusal to abandon racial discrimination and follow the world into a new era of greater equality and global integration. As a contrasting narrative, the following section provides a somewhat idealised vision of the kind of society South Africa might become if it pursues instead a broad programme to mitigate peak oil by adopting sustainability-oriented innovations across all sectors of the economy, as advocated in the foregoing chapters.

7.1 Vision of a Sustainable Post-oil Future

A positive vision of the future can serve as a general goal and guidepost for a society wanting to undertake a purposive transition towards (greater) sustainability. Hence we conduct a thought experiment: Based on currently known technologies, what might the South African socioeconomic system look like in 2050? This vision and its underlying norms are informed by earlier discussions and by the substantial literature penned by peak oil and sustainability writers (e.g. Greer 2009; Heinberg 2004; Hopkins 2008). It is also aligned with the vision statement for South Africa expressed in the National Framework for Sustainable Development:

> South Africa aspires to be a sustainable, economically prosperous and self-reliant nation state that safeguards its democracy by meeting the fundamental human needs of its people, by managing its limited ecological resources responsibly for current and future generations, and by advancing efficient and effective integrated planning and governance through national, regional and global collaboration. (DEAT 2008: 8)

Since energy is the master resource enabling all human activity, the appropriate point of departure for envisioning a sustainable society is its energy system, which will determine many other aspects of the broader socioeconomic system. By 2050, South African coal production will be about 30 years past its peak and much lower than it was in 2013. In addition, Sasol's CTL plant at Secunda reached the end of its functional economic lifespan more than a decade ago, and oil imports are negligible. Thus, to all intents and purposes, the oil age has drawn to a close. Instead, energy supply is derived largely from a diverse, localised, and decentralised mix of renewable energy sources. The western interior of the country is home to massive solar parks, including both concentrated solar thermal and photovoltaic (PV) farms. In addition, solar PV and solar water heaters are installed on virtually every residential and commercial building across the country. Geothermal heat pumps are also used extensively for space and water heating. Onshore and offshore wind farms generate electricity in the coastal areas of the Western Cape, Northern Cape, and Eastern Cape Provinces. Ocean current energy has more recently become an economically viable source of base-load power. The large-scale hydroelectric power plants have been maintained as before but are now complemented by thousands of micro-hydro turbines on small rivers and streams wherever these flow perennially close to human settlements and particularly on farms. These decentralised RE-generating units are connected by smart grids which manage the intermittency problem of solar and wind power as well as regulate demand to match available supply. Waste biomass is used to generate process heat, and cogeneration is used extensively in industrial processes. Nevertheless, society now has a much lower energy metabolism compared to the height of the fossil fuel era in the 2010s. Energy use has fallen substantially both per capita and per unit of economic activity, thanks to conservation and improved efficiency of energy generation and consumption.

The transport system has undergone a similarly radical transformation. Motorised transport is almost entirely powered by electricity. Most intercity passenger and freight transport is by rail, although transport between port cities is partly by advanced sailing ships. Metropolitan and urban transport is a mix of non-motorised transport (walking and cycling), grid-connected public transport (commuter and light rail), and electric vehicles (buses and some cars). In rural areas, the predominant mode of passenger transport is non-motorised (walking, cycling, and animal power), supplemented by electric vehicles. Freight is moved largely by rail and trucks powered by biodiesel. Air travel, which is now very limited, is powered by third-generation biofuels made from cellulose and algae. Long-distance business travel has been replaced by telecommuting and teleconferencing. Car sharing clubs are common for rural travel and local tourism. While the average level of mobility amongst wealthier households has decreased relative to that experienced in the peak of the car age, it has improved substantially for the portion of the population that was previously unable to afford private vehicles. The changes in mobility have been accompanied by health, safety, and environmental benefits.

The post-industrial agricultural system in 2050 in some ways resembles pre-industrial agriculture, although it is more diverse and knowledge intensive. Agricultural production is completely organic, with bio-fertilisers and bio-pesticides manufactured on a local basis. Use of agricultural machinery has been reduced substantially, and remaining farming vehicles such as tractors and harvesters are now fueled by locally produced biodiesel; many farmers set aside a portion of their land for biofuel crop production. Average farm size has declined drastically since peaking around 2020. Farming cooperatives are now the norm, and the percentage of the South African population involved in farming has risen substantially. In addition, small-scale urban agriculture is practiced in cities all over the country. A majority of the population spends at least some of their time producing some of their own food in local food gardens. Diets have become more seasonal and local. Human and other organic wastes are completely recycled within the agro-ecological food production systems.

The economy is in a steady-state, dynamic equilibrium, conforming to the requirements for sustainability and on a scale that is within the region's carrying capacity. Rates of material and energy throughput are roughly constant over the longer term, although with minor short-term fluctuations depending on climatic conditions. Recycling of materials, especially minerals and metals, is pervasive, and virgin mining is very limited. Integrated industrial processes operate largely on closed-loop, zero-waste principles. Products are characterised by high quality, durability, and modular designs that facilitate easy replacement of parts and recycling. Qualitative improvements continue to occur through technological and social innovations. Economies are predominantly localised, and many products are manufactured efficiently on a small scale with local raw materials. Employee-owned firms and cooperatives are more common than privately owned companies. The monetary system consists of local currencies that are connected into larger regional and global monetary networks, facilitating trade at various scales, depending largely on energy efficiency coefficients.

The population of South Africa has shrunk somewhat from its peak in the second decade of the century, thanks to falling fertility rates following improved birth control and empowerment of women. Absolute poverty no longer occurs as communities work together to ensure that everyone's basic needs are met. The distribution of income and wealth is much more egalitarian, with the Gini coefficient having nearly halved from its 2010 level. Both jobs and work hours are shared more evenly, and a 4-day work week is standard. Political power is decentralised and based on a highly participatory system that is facilitated by Internet-based communication and voting tools. A variety of sustainable human settlements have emerged. Cities have become much more compact, with higher density core business and residential areas surrounded by "urban villages". Eco-village-type settlements with sustainable building designs and small footprints have sprung up in rural areas with good arable land and favourable climate conditions. The industrial-era materialist and competitive value system has been replaced by a new ethic that embraces diversity, sharing, cooperation, self-sufficiency, and local resilience.

These socioeconomic subsystems exhibit some common features. They are all characterised by decentralised power systems (energy, economic, political) and strong network relationships. Linear flows of materials and waste have been replaced by circular flows involving reuse and recycling. Respect for ecosystems as the basis of human welfare is an integral element of the new culture.

7.2   A Transition Action Plan

Having described an idealistic vision of the future, the next question is the following: "How do we get there?" As it happens, a national strategy and action plan for transitioning South Africa to a sustainable society have already been developed, in the form of the National Strategy for Sustainable Development (NSSD). To date, however, it has yet to be implemented in practice as the overarching policy strategy by national government. In fact, it has largely been ignored—probably because it was developed by the Department of Environmental Affairs and not one of the politically stronger "economic cluster" departments. Moreover, the NSSD does not give sufficient attention to planning for oil depletion, which should be one of the key strategic priorities. This section therefore addresses the aspects of a sustainability transition action plan that are geared towards systematically reducing the oil dependence of the socioeconomic system, aiming for complete oil independence within about 40 years, but without sacrificing other developmental goals. It is written in the spirit of Howard and Elizabeth Odum's book *A Prosperous Way Down* (Odum and Odum 2011).

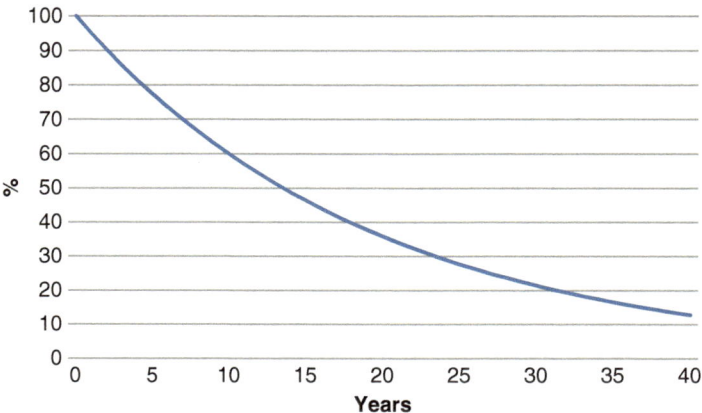

Fig. 7.1 Targeted cumulative reduction in petroleum consumption (5 % per year). *Source*: Author's calculations. *Note*: The curve above represents an assumed 5 % per annum decline in fuel consumption in order to illustrate the cumulative reduction after 5-yearly intervals

In my opinion the National Planning Commission (NPC) should take primary responsibility for co-ordinating the oil transition plan. It should appoint a National Oil Independence Task Team comprising a relatively small group of experts. This Task Team would deliver recommendations for deliberation by a larger caucus comprising representatives from all national government departments, organised business and labour, and civil society organisations. Through a consultative process, a National Transition Action Plan (NTAP) would be formulated. The oil independence plan must be consistent with the government's National Climate Change Response Policy and be integrated with the NPC's National Vision and Strategic Plan. In addition, due to geographic specificities, individual cities, towns, and regional districts should formulate their own local transition action plans (in accordance with the NTAP), based on collaboration between municipal governments and local communities and businesses and included in municipal-level Integrated Development Plans.

The NTAP should adopt a specific target for annual reductions in domestic petroleum consumption. A 5 % per annum reduction seems a reasonable goal, based on a mid-range estimate of the rate at which world oil exports will contract (see Chap. 1) and assuming that the lifespans of Sasol's Secunda CTL plant and PetroSA's Mossgas GTL plant will have ended by 2050, having been commissioned in 1982 and 1993, respectively. This rate of decline would require a cumulative 50 % reduction after 14 years and an 80 % reduction after 30 years (see Fig. 7.1). In reality, of course, fuel savings are unlikely to decline smoothly, as some measures require lumpy investments (e.g. for transport modal shifts from road to rail).

As discussed earlier (see Figs. 2.6 and 2.9), both the absolute and per capita consumption of petroleum products in South Africa has been on a rising trend over the past two decades, albeit with a significant dip in the recession of 2008–2009. The challenge for the forthcoming era of declining global oil availability is to reduce oil consumption both in absolute terms and per unit of GDP while continuing to develop the economy and reduce poverty, unemployment, and inequality. One vital dimension of reducing total oil consumption over the long term is to curtail population growth; otherwise, the demands of an expanding population will offset many of the efficiency gains recommended in this book and will mean that scarce resources have to be spread more thinly amongst the populace. The empowerment and education of women, together with the provision of family planning services, can make important contributions towards stabilising the size of the population (Brown 2008). In addition, once the availability of world oil exports begins to shrink and world oil prices rise in response, a much more rapid degree of decoupling between petroleum consumption and real GDP will be required than the 1.4 % per annum average decoupling that was achieved between 1998 and 2012. If petroleum consumption falls by 5 % per annum, then the rate of relative decoupling would need to rise to 5 % per annum just to maintain real GDP at a constant level. In what follows, the major interventions that are required in the short, medium, and long term are summarised, based on the detailed policy recommendations presented in Chaps. 2–5.

Short-term planning and implementation (i.e. 1–2 years) should focus on two major areas. The first step should be a national education and awareness programme to properly inform all sectors of society of the need to reduce petroleum consumption and become more sustainable in their lifestyles and business operations. This could take the form of information campaigns on national television and radio stations as well as the press and online social media. It is essential for the government to mobilise popular support for the transition plan, as it has attempted to do with its National Development Plan.

Secondly, government should formulate emergency response plans to deal with sudden and drastic limitations on petroleum imports. Probably the largest short-term risk is a broad military conflict in the Middle East, perhaps sparked by the Syrian civil war and its spill-over to neighbouring countries. This could potentially lead to the closure of the Straits of Hormuz in the Persian Gulf, through which 17 million barrels of oil per day transited in 2011, representing 35 % of world seaborne oil and almost 20 % of oil traded globally (EIA 2012). South Africa sources approximately a half of its crude oil imports from Middle Eastern suppliers, so a complete interruption of these supplies could reduce South Africa's total petroleum fuel supply by about a third. Thus national government should formulate plans to cope with the logistical implications of a sudden loss of a third of the country's fuel supply, which could potentially last several months. Authorities must be ready to implement measures for rapid demand restraint in combination with a fuel rationing system. The rationing system must include a fuel prioritisation plan to ensure the continued functioning of essential services (police, defence force, fire, rescue, and ambulance services), transport of medical supplies, and food production and distribution. In addition, a food security plan should incorporate a warehousing system to make provision for disruptions to the just-in-time food delivery system. Given the limitations that refineries have on the proportions of various petroleum products that they can extract from a barrel of crude oil, consumption of diesel and petrol would have to shrink by similar percentages, implying that both passenger and freight transport will be affected. The greatest area for fuel saving with the least negative socioeconomic impacts would be (1) discretionary driving (e.g. for leisure) and motor racing; (2) single-occupant commuting; and (3) freight movement along corridors where excess rail capacity exists. Air travel could be drastically reduced since it is mostly undertaken for leisure and for business commuting that could be replaced by telecommuting, although the air freight business would be negatively affected. Finally, the state must have plans in place to respond to social panic, hoarding behaviour, or civil unrest that could accompany dislocations in fuel and food supplies and steeply rising prices.

Medium-term oil independence initiatives should aim to alter the behaviour of consumers and producers. This can be achieved firstly through the introduction of regulations for energy (liquid fuel) conservation and efficiency, car-pooling (e.g. via dedicated car pool lanes), and vehicle fuel efficiency standards. Secondly, fiscal incentives (e.g. taxes, subsidies, and rebates) should be introduced to promote the manufacture and purchase of bicycles, electric bikes and scooters, battery electric vehicles, and plug-in hybrid vehicles; dissuade purchases of internal combustion engine vehicles; and to encourage modal shifts from private to public transport and from road to rail freight.

Long-term strategies for reducing oil dependence centre on new infrastructure investments for sustainable energy and transport. Specific infrastructure investment projects will need to be informed by detailed studies based on appropriate evaluative methodologies, such as life cycle analysis, cost–benefit analysis, and/or cost-effectiveness analysis. Key criteria for energy investments should be the energy return on investment and net energy delivered, while for transport investments the oil saved per unit of energy invested is a crucial variable. Perhaps any new transport infrastructure with an intended lifespan of over 25 years should only be built if it is compatible with a zero-oil future. Thus, for example, government- and state-owned enterprises might consider light rail trains or trams rather than bus rapid transit systems, unless the buses could ultimately be powered by electricity. Another critical question, which will need to be answered on a geographic specific basis, is whether to focus on providing non-oil-based transport to ensure that people can access services and work opportunities or rather to achieve this by channelling resources into relocating people and redesigning human settlements. In other words, is it more resource and cost efficient to move people on a continual basis or to move physical capital once-off? Similar questions pertain to freight transport. Another crucial aspect of long-term planning is the development of appropriate skills. This will require the formulation and promotion of new curricula for primary, secondary, and tertiary education institutions that are compatible with an era of declining oil availability, such as agro-ecological farming, engineering skills for renewable energy, and sustainable transport technologies.

Finally, the national and local transition action plans should be subject to annual monitoring and evaluation processes, which will evaluate progress made towards reaching the petroleum conservation targets. The appropriateness of the targets themselves should also be reviewed on an annual basis and revised if necessary depending on global and national circumstances.

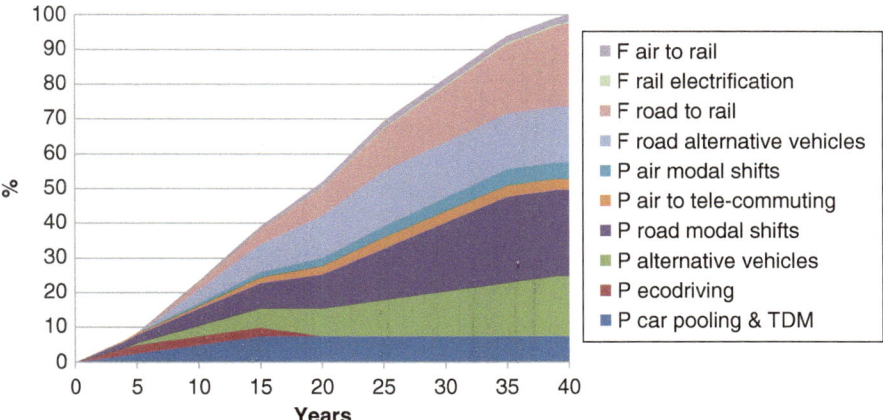

Fig. 7.2 Potential fuel savings in the transport sector. *Source*: Author's calculations. *Notes*: *P* passenger, *F* freight, *TDM* travel demand management. Each of the "wedges" represents the percentage of 2010 transport fuels that could be saved through the implementation of groups of fuel saving measures. The various wedges are added vertically to indicate the potential total fuel savings. Some interventions will take longer to implement than others. Savings from eco-driving are assumed to phase out as the current internal combustion engine vehicle fleet is replaced by electric-powered vehicles and mass transit. For details of assumptions and calculations, see Appendix G in Wakeford (2012)

To conclude this section, consideration is given to the sectoral components of petroleum product demand and how these can be reduced through conservation and substitution over a 40-year time horizon. As noted in Chap. 2, the sectoral shares of oil product demand in 2010 were as follows: transport (69.8 %), agriculture (4.4 %), industry (4.3 %), commercial and public services (3.8 %), residential (2.7 %), and non-energy uses (14.9 %) (IEA 2013). Each sector will need its own oil independence action plan. For agriculture, the goal should be to reduce liquid fuel consumption (e.g. by employing more labour-intensive production methods), perhaps by about half, and to substitute the remaining fuel consumption with locally produced biodiesel. Industry uses petroleum products mostly in the form of lubricants and diesel to run machines and small-scale electric generators. The latter will need to be substituted with renewable electricity, and innovation will be required to develop lubricants from biomass. The residential sector and commercial and public services use petroleum mainly in the form of LPG, paraffin, and diesel for small electric generators, all of which need to be replaced by renewable energy electricity, biogas, and biomass over time. For the transport sector, a scenario for achieving 100 % fuel savings after 40 years through a range of mitigation measures is illustrated in Fig. 7.2. Fuel savings are expressed as a percentage of the petroleum fuels consumed by passenger (road and air) and freight (road, rail, and air) transport in 2010. While deviating slightly from the 5 % per annum target, this scenario is deemed plausible assuming that the economy remains intact and the will to formulate and implement policies is forthcoming.

7.3   Barriers and Risks to Implementation

Numerous obstacles and risks might potentially thwart the implementation of the transition strategy. Recognising that there are overlaps between them, seven categories of barriers to transition can be identified, namely, cultural–ideological, behavioural–psychological, social, political, institutional, economic, and environmental (van den Bergh et al. 2011).

At an ideological level, adherence to an outdated intellectual paradigm (such as neoclassical economics that ignores the central importance of energy and resources to economic systems) or cornucopian faith in the ability of technology to overcome resource constraints could lead to a failure to recognise the challenge posed by peak oil and a continuation along the path of denial and false hopes. Ideological lock-in would likely result in a misinterpretation of economic signals, for instance financial volatility and economic distress not being understood as symptoms of underlying energy constraints. It appears that the prevailing ideological and policy paradigm in South Africa will have to undergo a fundamental shift if a proactive transition action plan is to be adopted and implemented.

Similarly, the persistence of cultural values such as consumerism, greed, individualism, and independence would result in citizens attempting to cling to current or aspired-to consumer lifestyles and preferences (Geels 2011: 25). For example, continued consumer preferences for ICEVs, which historically have offered a high degree of mobility, could thwart the introduction of more sustainable alternatives such as car-pooling initiatives and use of public transport and non-motorised transport. Cultural conservatism could hinder the kinds of social innovations that are required for a sustainability transition. Behavioural–psychological traits such as bounded rationality and short time horizons (myopia) amongst producers and consumers could also obstruct appropriate adaptive responses (van den Bergh et al. 2011: 7). Citizens could react to socioeconomic challenges by blaming particular sectors of society or the government for crises, instead of cooperating to find solutions. Socially, a serious lack of education, flexible skills, and an entrepreneurial culture that are required to support innovation and the adoption of new technologies is a constraint of particular relevance in a developing country like South Africa.

Political barriers to the implementation of a transition strategy involve power relations among various groups in the country. Lobbying by incumbent vested interests serves to preserve the status quo and militates against the adoption of new policies (Barbier 2011). In South Africa there are several vested interests that wield considerable political power. The most important is the minerals-energy complex (MEC), a component of which is the "energy-intensive users group", which comprises the largest industrial consumers of electricity and which apparently had a significant influence on the development of the *Integrated Resource Plan for Electricity* (Baker 2011). Another member of the MEC is the electricity utility, Eskom, which by virtue of its position as monopoly purchaser and distributor of electricity has—at least until very recently—stymied the attempts of independent power producers to supply power to the national grid (Baker 2011; Trollip and Tyler

2011). Sasol, also a key component of the MEC, apparently used its economic power to influence the National Treasury's decision in 2006 not to impose a windfall tax on synthetic fuels by intimating that it would cancel its plans to build a new coal-to-liquid plant (National Treasury 2007). If this "Project Mafutha" did materialise, it would further entrench lock-in to the fossil fuel-ICEV regime nexus and retard South Africa's transition to a sustainable socioeconomic regime. Similarly, the state-owned oil company PetroSA is intent on building a large new oil refinery at the port of Coega in the Eastern Cape, based on the assumption that domestic and regional petroleum demand will continue to grow for decades. Should this "Project Mthombo" be approved by Cabinet, it could represent a monumental misallocation of resources that the country can ill afford.

A second group that stands in opposition to oil mitigation policies is those companies with an interest in maintaining the dominance of road-based transport. This group includes automobile manufacturers, who would oppose the withdrawal of state subsidies for ICEV production, likely using potential job losses as a lever. Other key members of this interest group are the South African National Roads Agency Limited (SANRAL) and the road construction companies, which profit from upgrades and extensions to national and urban paved roads. Third, trade unions might resist support for new sectors such as manufacture of renewable energy technologies to the extent that this would require a shift of resources away from established industries. Fourth, financial institutions (dominated by an oligopoly of five banks) are likely to vigorously oppose proposals for monetary reform, as they would stand to lose some of their considerable profits.

Overcoming these vested interests is a tall order, although to some extent it might be made easier by unfolding economic circumstances, which will erode the financial power of resource- and energy-intensive sectors and commercial banks. For example, the MEC is "under threat from rising coal costs, electricity supply issues, rising tariffs and a utility struggling to hold onto its monopoly" (Baker 2011: 28). Similarly, the power of commercial banks is being weakened in other countries by the debt crisis, and this could spill over to South Africa. Nevertheless, the fact that "[t]here is no one national champion of renewable energy to forge a path but rather a number of different entities acting at times in competition with each other" (Baker 2011: 28) makes it harder for the RE niches to challenge the dominance of incumbent fossil fuel regimes.

Many policy and institutional failures result from institutional inertia—the persistence over time of institutions (in the sense of formal and informal social rules governing human behaviour), which have often evolved to minimise transaction costs (Barbier 2011). The social order that evolved with a given resource base (e.g. fossil fuels) locks in that development trajectory, since it may be incompatible with reducing transaction costs associated with an alternative resource base (Barbier 2011: 61). More specific institutional constraints include government failure resulting from rent-seeking behaviour, corruption in state apparatuses (which could deteriorate as people in positions of power attempt to cling to their wealth as the economy deteriorates), and a lack of institutional and managerial capacity in the state. Furthermore, the capacity of the state to implement the transition strategy might be eroded by the impacts of declining world oil production, such as tightening financial constraints and escalating social protests that undermine the state's capacity to govern. Heinberg (2004) points out that his "powerdown" scenario presents a paradox to national governments, in that they need to implement policies that in the long term will undermine their own power. However, if they fail to implement the transition strategy, their capacity to govern will be steadily eroded in any case—and probably much more rapidly. Yet another example of institutional failure is a lack of coordination among state-owned enterprises and agencies, such as disputes between railway operator Transnet and the Passenger Rail Authority of South Africa (PRASA) (Fin24 2011). Lastly, some aspects of a transition strategy will be circumscribed to a degree by global institutional factors such as the World Trade Organisation (WTO), the global financial architecture, and international climate negotiations.

Economic barriers to a sustainability transition include various kinds of market failures as well as unfavourable financial conditions. The first market failure is that of imperfect information: most economic agents are grossly ill informed about the likely future trajectory of global oil production and the risks of continued oil dependence. This is compounded by explicit and implicit subsidies for fossil fuel consumption and production, which hinder improvements in energy efficiency and the growth of renewable energy sources (Barbier 2011: 64). Second, sustainability is a public good which is subject to free-rider and prisoner's dilemma problems (Geels 2011: 25). Third, there are several externalities, both positive and negative, associated with sustainability-oriented innovations (van den Bergh et al. 2011: 4–5). One is negative environmental externalities, which result in socially damaging investments being made. Another is positive knowledge/information externalities, which lead to suboptimal levels of private investment in research and development and innovations. A third is lock-in to existing (unsustainable) socio-technical regimes, which can occur as a result of economies of scale, sunk investments in infrastructure and skills, land tenure systems, actions by professional and fraternal organisations to maintain the status quo, and various incentives and regulations (Geels 2011: 25; Westley et al. 2011: 34). For example, the MEC favours the construction of new coal power plants, possibly another coal-to-liquid plant, and the extraction of shale gas. The question, which needs detailed investigation, is whether these preferences are merely exhibiting lock-in to a fossil fuel regime or whether it is based fundamentally on higher energy return on investment ratios for fossil fuels compared to renewable energy sources. In addition, financial constraints will almost certainly pose a major obstacle to transition, for instance in the form of a domestic credit crunch and cost escalation, especially for mega-projects such as new railways.

Finally, environmental factors may hinder the implementation of certain preferred elements of a transition strategy. These may take the form of resource depletion or scarcity (e.g. related to water, or rare earth metals that are required for many renewable energy technologies such as neodymium for wind turbines or lithium for batteries) or environmental impacts of new technologies or climate change impacts (e.g. upon smallholder farming yields).

In addition to these myriad obstacles, the implementation of a transition strategy faces various risks. The first risk is that the transition strategy is a political non-starter. Proactive leadership might not be forthcoming, with no champion emerging to tackle this issue as a national emergency. Instead, the state could respond with reactive crisis management in which short-term solutions are favoured for expediency rather than long-term efficiency, locking the country into decisions that will compound the problems in the longer term. The evidence and argument presented in this book aim to reduce this risk.

The second set of risks involves unintended consequences of transition policies and measures. For example, government acknowledgement of peak oil might result in a collapse in business and consumer confidence and a resulting crash in the stock market and contraction of economic activity. However, in 2006 the Swedish Government publicly adopted a plan to reduce oil dependency by 50 % by 2020, and this did not result in substantial adverse impacts on the country's economy or stock market. Another common worry is that the introduction of unconventional policies, such as monetary reform, could trigger rapid capital flight. On the other hand, capital flight could be substantially worse later on if proactive mitigation strategies are not implemented and the economy deteriorates more rapidly. Concerns over the rebound effect of energy efficiency measures are much less significant in a context of rising energy prices. There may be substantial labour unrest as state support undergoes sectoral shifts; but significant sectoral economic shifts will take place in any case. While it is not possible to foresee all potential unintended consequences, this should not hamstring the transition strategy as a whole.

A third category of risks may be termed "black swan" events (following Taleb 2007). These include severe oil price shocks (e.g. rising to over $200 per barrel) and shortages (e.g. a 10 % loss of world oil output) occurring soon and bringing severe economic consequences, such as a collapse of the global financial system; a world war involving a majority of the major powers, for example centred on the oil-rich Middle East and Central Asia; invasion of South Africa by a foreign power in order to gain control over its strategic mineral resources (e.g. coal, uranium, platinum group metals, magnesium, and chromium); regional immigration on a scale that overwhelms the capacity of the state; and massive regional climate change impacts resulting in widespread crop failures, food shortages, and social chaos. Any of these events would dramatically worsen the prospects for a successful transition, at least in the short to medium term. However, prior efforts towards transition would likely mitigate their impacts in any event, so if anything the possibility of black swans increases the motivation for early transition initiatives.

A fourth type of risk could be termed a "pink swan". An example would be a major discovery of oil or natural gas within South Africa's territorial boundaries (including offshore areas), which could be a game changer if it was sufficiently large to replace imported oil. However, pricing would still be an issue, as the oil or the gas would likely be extracted at least partially by foreign-owned private companies that would seek to sell the hydrocarbons for the highest price attainable in the international market. The only way such a discovery would make a material difference to South Africans would be if the state (possibly through its oil company

PetroSA) negotiated sufficiently large royalties that it would be in a financial position to subsidise domestic petroleum products. Should such a "pink swan" materialise, however, the rents accruing from the resource should be diverted to sustainability-enhancing investments.

On the other hand, there is also the possibility of what might be termed "green swans", namely, disruptive technological developments that render older investments obsolete. Examples might include revolutionary developments in solar cell efficiency, electricity storage technology, commercial production of biodiesel from algae on a large scale, success in the quest to commercialise low-energy nuclear reactions (formerly termed "cold fusion"), and radically new propulsion mechanisms for transport. The risk posed by such green swans is that early adoption of current, less efficient and less effective technologies would result in lock-in and reduce the potential for widespread deployment of the newer technologies. Nonetheless, any of these new technologies would likely take a decade or more to be commercialised and scaled up to a significant extent, whereas transition planning should ideally begin immediately to mitigate the risks of early-onset impacts from peaking world oil production. The possibility of "green swans" does however suggest that a transition strategy should be designed as flexibly as possible so that new, disruptive technologies can be taken advantage of if or when they arise.

In a similar vein, there is a risk that elements of the transition strategy follow dead-end paths. This could occur if the uncertainties surrounding the timing of oil shock impacts and behavioural responses lead to a misallocation of resources, for example if the state were to build new public transport infrastructure in cities, only to have a large exodus of people to rural areas in later years. Other dead ends could be reached for energy, transport, and housing infrastructure projects that cannot be completed, for instance if cost escalation (driven for instance by increasing global demand for mitigation-related capital equipment, higher energy and material input costs, and rising interest rates) intersects with shrinking tax revenues to render the projects unaffordable. Again, steering investments in a more sustainable manner from now on will attenuate this risk.

The challenge at this point is that insufficient information is available on some of the key energy-related issues that are vital to achieving greater sustainability. There is a risk that government and society will make decisions based on insufficient analyses that superficially appear sound (perhaps because they conform to global trends) but which in reality are not. Examples include the massive but probably at best energy-neutral US programme of corn-based ethanol and possibly the subsidisation of solar PV systems which recent research suggests may deliver low net energy yields. Would a continuation of urbanisation with urban densification and urban agriculture be better energetically and economically than effective rural development that includes measures to boost smallholder agroecological farming? There is simply not enough detailed information on the sustainability-related aspects of both energy and transport infrastructure projects to guide policy makers in South Africa. Moreover, we do not fully understand the extent to which non-fossil energy sources (both renewables and nuclear power) and electricity-driven forms of transport depend fundamentally on a fossil fuel-based economic system.

Thus what is required is a research programme that undertakes a comprehensive, system-oriented analysis of the costs and benefits of each alternative, utilising methodologies such as energy input–output analysis and life cycle analysis. Such analyses should include economic, environmental, and social costs and benefits over the full life cycle of alternative technologies. As Charles Hall and colleagues have argued, the prosperity of an economy—and hence a society—depends not simply on having sufficient energy per se but on having enough good-quality energy with a very significant EROI (Hall et al. 2009; Hall 2011). Therefore the EROI ratios of domestic energy alternatives need to be compared with one another and with the EROI of imported oil. Perhaps the most important recommendation emanating from this book is that there is an urgent and compelling need for a well-funded, independent national institute to study these issues in depth.

7.4 Beyond South Africa

The significance of this national case study research goes beyond the geographic borders of South Africa. As argued in the introduction, South Africa in many ways represents a microcosm of the world at large. It has an industrial economy with sophisticated agribusiness, mining, petrochemical, manufacturing, and financial service industries, while nearly half the population lives in poverty and at least a quarter suffer from food insecurity. The country comprises many different ethnic groups and has a high degree of income and wealth inequality. Furthermore, South Africa is confronted by most of the same environmental challenges that face the world as a whole, such as fossil fuel dependence, water scarcity, and climate change. As in other countries, there are powerful vested interests, especially in the fossil fuel sector and its affiliates, which actively work to maintain the status quo. Thus the challenges posed by peak oil, as well as the mitigation responses that are called for, are broadly similar for South Africa and for the world as a whole. Both need to implement policies and measures to decouple economic well-being from oil consumption and to insulate themselves from the impacts of oil shocks.

Moreover, most of the specific mitigation policies and measures advocated in the preceding chapters are applicable to other net oil-importing countries, both industrialised and developing, to a greater or a lesser degree depending on their economic structures and energy endowments. It is clear that industrialised countries cannot sustain their current fossil fuel-intensive socio-metabolic rate for much longer. In fact consumption of fossil fuels—including oil—has already been declining in most OECD countries since 2007, and their economies have stopped growing (virtually or actually). But given their continued high degree of oil dependence, most major systems (e.g. energy, transport, agriculture, petrochemicals) will have to undergo a profound transition over the coming decades. For their part, developing countries cannot continue to follow the fossil fuel-intensive path of industrial development because there are insufficient quantities of high-quality, high-EROI, affordable fossil fuel resources remaining. The challenge for these countries is to leapfrog to a much more efficient and sustainable socio-metabolic regime that is based largely on renewable energy resources.

South Africa surprised the world in 1994 when it achieved a largely peaceful democratic transition after centuries of colonialist exploitation and race-based minority rule. The country also became the first nation in the world to unilaterally disarm all its nuclear weapons in the early 1990s. Can South Africa rise to the challenge of peak oil and lead the world in a transition to a sustainable future? The resources and technologies are available; what are needed above all are sufficient awareness, political will, enlightened leadership, and individual resolve. Future generations are depending on us for their well-being—we owe it to them to give it our best shot.

References

African Loft (2008) South Africa WIZZIT: banking the unbanked with mobile phones. African Loft Media. http://www.africanloft.com/south-africa-wizzit-banking-the-unbanked-with-mobile-phones/. Accessed 20 Jun 2011

Aleklett K, Campbell CJ (2003) The peak and decline of world oil and gas production. Miner Energ 18(1):5–20

Aleklett K, Hook M, Jakobsson K, Lardelli M, Snowden S, Soderbergh B (2010) The peak of the oil age—analyzing the world oil production reference scenario in world energy outlook 2008. Energ Pol 38(3):1398–1414

Aliber M, Hart TG (2009) Should subsistence agriculture be supported as a strategy to address rural food insecurity? Agrekon 48(4):434–458

Allen W, Wilson C (2012) The good food revolution: growing healthy food, people and communities. Penguin, New York

Altieri MA (2009) Agroecology, small farms and food sovereignty. Mon Rev 61(3):102–113

Amigun B, Sigamoney R, von Blottnitz H (2008) Commercialisation of biofuel industry in Africa: a review. Renew Sustain Energ Rev 12(3):690–711

APPGOPO (2009) Tradable energy quotas (TEQs): a policy framework for peak oil and climate change. All Party Parliamentary Group on Peak Oil, UK House of Commons, London

ASPO Ireland (2009) Newsletter no. 100. Association for the Study of Peak Oil—Ireland, Dublin

ASPO-SA, Vanderschuren M, Lane T (2008a) Energy conclusions and recommendations for NATMAP. Research report submitted to the National Department of Transport. Association for the Study of Peak Oil—South Africa, Cape Town

ASPO-SA, Vanderschuren M, Lane TE (2008b) Energy and transport status quo. Research report submitted to the National Department of Transport. Association for the Study of Peak Oil—South Africa, Cape Town

ASPO-SA, Vanderschuren M, Lane TE (2008c) Reducing oil dependency and alternatives to oil-based liquid fuel transport. Research report submitted to the National Department of Transport. Association for the Study of Peak Oil—South Africa, Cape Town

ASPO-SA, Vanderschuren M, Lane TE (2008d) Transport scenarios under peak oil—to 2030. Research report submitted to the National Department of Transport. Association for the Study of Peak Oil—South Africa, Cape Town

Baiphethi MN, Jacobs PT (2009) The contribution of subsistence farming to food security in South Africa. Agrekon 48(4):459–482

Baker L (2011) Governing electricity in South Africa: wind, coal and power struggles. Paper presented at the 10th Nordic environmental science conference, Stockholm University, Stockholm

Barbier EB (2009) A global green new deal. Report prepared for the economics and trade branch, division of technology, industry and economics, United Nations Environment Programme

Barbier EB (2011) Transaction costs and the transition to environmentally sustainable development. Environ Innov Societal Transitions 1(1):58–69

Benes J, Chauvet M, Kamenik O, Kumhof M, Laxton D, Mursula S, Selody J (2012) The future of oil: geology versus technology. IMF Working Paper WP12/109. International Monetary Fund, Washington, DC

Berkhout PHG, Muskens C, Velthuijsen JW (2000) Defining the rebound effect. Energ Pol 28(6–7): 425–432

Berman AE (2010) Shale gas—abundance or mirage? Why the Marcellus shale will disappoint expectations. The Oil Drum. http://www.theoildrum.com/node/7075. Accessed 15 May 2011

Biodiesel Centre (2011) Biodiesel centre. http://www.biodieselcentre.co.za/html/full_systems. html. Accessed 1 Mar 2011

BP (2013) Statistical review of world energy 2013. BP plc, London

Brown LR (2008) Plan B 3.0: mobilizing to save civilization. W.W. Norton & Company, New York/London

Brown LR (2012) Full planet, empty plates: the new geopolitics of food security. W.W. Norton & Company, New York

Brown JJ (2013) Comment: the export capacity index. Post Carbon Institute. http://www.resilience. org/stories/2013-02-18/commentary-the-export-capacity-index. Accessed 1 Mar 2013

Burkhardt P (2013) PetroSA plans South Africa LNG terminal of up to $510 million. Bloomberg Businessweek. http://www.businessweek.com/news/2013-07-18/petrosa-plans-south-africa-lng-terminal-of-up-to-510-million. Accessed 23 Aug 2013

Bushnell D (2012) Low energy nuclear reactions, the realism and the outlook. NASA. http://futureinnovation.larc.nasa.gov/view/articles/futurism/bushnell/low-energy-nuclear-reactions.html. Accessed 4 Mar 2013

Campbell CJ (2006) The Rimini protocol an oil depletion protocol: heading off economic chaos and political conflict during the second half of the age of oil. Energ Pol 34(12):1319–1325

CES (2013) Community exchange system. http://www.ces.org.za. Accessed 1 Mar 2013

Chakauya E, Beyene G, Chikwamba RK (2009) Food production needs fuel too: perspectives on the impact of biofuels in southern Africa. S Afr J Sci 105(5):174–181

City of Portland (2007) Descending the oil peak: navigating the transition from oil and natural gas. City of Portland Peak Oil Task Force, Portland, OR

Cleveland CJ (2008) Energy return on investment (EROI). In: Cleveland CJ (ed) The encyclopedia of earth. Environmental Information Coalition, National Council for Science and the Environment, Washington, DC

Creutzig F, Papson A, Schipper L, Kammen DM (2009) Economic and environmental evaluation of compressed-air cars. Environ Res Lett 4(4):1–10

CSIR (2010) 6th annual state of logistics survey for South Africa 2009. Council for Scientific and Industrial Research, Pretoria

CSIR (2013) 9th annual state of logistics survey for South Africa 2012. Council for Scientific and Industrial Research, Pretoria

Curtis F (2003) Eco-localism and sustainability. Ecol Econ 46:83–102

Curtis F (2009) Peak globalization: climate change, oil depletion and global trade. Ecol Econ 69:427–434

DAFF (2012) Abstract of agricultural statistics 2012. Department of Agriculture, Forestry and Fishing, Pretoria

DAFF (2013) Economic review of the South African agriculture 2012. Department of Agriculture, Forestry and Fishing, Pretoria

Davenport J (2010) New fund considered to tackle R75bn road maintenance backlog. Engineering News. http://www.engineeringnews.co.za/article/new-fund-considered-to-tackle-r75bn-road-maintenance-backlog-2010-04-13. Accessed 13 Apr 2010

DCoGTA (2011a) Local economic development. Department of Cooperative Governance and Traditional Affairs. http://www.thedplg.gov.za/subwebsites/led/index.html. Accessed 16 May 2011

DCoGTA (2011b) About the CWP. Department of Cooperative Governance. http://www.cogta.gov.za/cwp/index.php/about-the-cwp.html. Accessed 16 May 2011

DEA (2010) National strategy and action plan for sustainable development. Department of Environmental Affairs, Pretoria

DEAT (2008) National framework for sustainable development. Department of Environmental Affairs and Tourism, Pretoria

Diamond J (2005) Collapse: how societies choose to fail or succeed. Viking, New York

Dittmar M (2013) The end of cheap uranium. Sci Total Environ 461–462:792–798

DME (2007a) Energy security master plan—liquid fuels. Department of Minerals and Energy, Pretoria

DME (2007b) Biofuels industrial strategy. Department of Minerals and Energy, Pretoria

DMR (2012) Report on investigation of hydraulic fracturing in the Karoo basin of South Africa. Department of Mineral Resources, Pretoria

DoE (2011) Electricity regulations on the integrated resource plan 2010–2030. South African Government Gazette No. 9531. Department of Energy, Pretoria

DoE (2012) Energy balances for South Africa. Department of Energy. http://www.energy.gov.za/files/media/media_energy_balances.html. Accessed 25 Feb 2013

Donnelly L (2010) Synfuels to make up shortfall? Mail & Guardian 3 September

Donnelly L (2011) Car makers cash in on subsidies. Mail & Guardian Online. http://mg.co.za/article/2011-05-30-car-makers-cash-in-on-subsidies. Accessed 12 Oct 2011

DoT (2005) Key results of the national household travel survey 2003. Department of Transport, Pretoria

DoT (2008a) National transport master plan 2005–2050. Status quo, vol 1. Department of Transport, Pretoria

DoT (2008b) Draft national non-motorised transport policy. Department of Transport, Pretoria

Douthwaite R (1996) Short circuit: strengthening local economics for security in an unstable world. Lilliput Press, Dublin

Douthwaite R (2010) The supply of money in an energy-scarce world. In: Douthwaite R, Fallon G (eds) Fleeing vesuvius. FEASTA, Dublin, pp 58–83

DPLG (2006) National framework for local economic development in South Africa. Department of Provincial and Local Government, Pretoria

DPW (2011) Expanded public works programme. Department of Public Works. http://www.epwp.gov.za/. Accessed 19 May 2011

DRDLR (2009) The comprehensive rural development programme framework. Department of Rural Development and Land Reform, Pretoria

DTI (2011) Industrial policy action plan 2011/12–13/4. Department of Trade and Industry, Pretoria

DTI (2012) South African Trade Statistics. Department of Trade and Industry. http://www.thedti.gov.za/econdb/raportt/defaultrap.asp. Accessed 15 Jan 2013

Eberhard A (2011) The future of South African coal: market, investment and policy challenges. Stanford program on energy and sustainable development, working paper #100. Stanford University, San Francisco

EDD (2010) The new growth path: framework. Economic Development Department, Pretoria

EIA (2011) World shale gas resources: an initial assessment of 14 regions outside the United States. US Energy Information Administration, Washington, DC

EIA (2012) World oil transit chokepoints. US Energy Information Administration. http://www.eia.gov/countries/regions-topics.cfm?fips=WOTC. Accessed 10 Sept 2013

EIA (2013a) International energy statistics. US Energy Information Administration. http://www.eia.doe.gov/emeu/international/contents.html. Accessed 20 Aug 2013

EIA (2013b) Technically recoverable shale oil and shale gas resources: an assessment of 137 shale formations in 41 countries outside the United States. US Energy Information Administration, Washington, DC

EIA (2013c) South Africa country analysis brief. US Energy Information Administration, Washington, DC eNatis (2013) Live vehicle population. National Traffic Information Service. http://www.enatis. com. Accessed 25 Feb 2013

Engineering News (2011) Cost of delayed Nigeria GTL plant now $8,4bn 25 Feb

Eskom (2010) Underground coal gasification. Eskom. http://www.eskom.co.za/live/content. php?Item_ID=14077. Accessed 18 Feb 2011

Essama-Nssah B, Go D, Kearney M, Korman V, Robinson S, Thierfelder K (2007) Economywide and distributional impacts of an oil price shock on the South African economy. Policy research working paper 4354. World Bank, Washington, DC

FAO (2005) Fertilizer use by crop in South Africa. Food and Agriculture Organisation, Rome

FAO (2010) Conservation agriculture. United Nations Food and Agriculture Organistation, Rome

Feasta (2005) South Africa and the oil price crisis. Foundation for the Economics of Sustainability, Dublin

Feasta (2008) Cap and share: a fair way to cut greenhouse gas emissions. Foundation for the Economics of Sustainability, Dublin

Fin24 (2011) PRASA wants mediator for transnet debt row. Naspers. http://www.fin24.com/ Companies/PRASA-proposes-mediator-for-Transnet-debt-row-20101018. Accessed 7 Nov 2011

Fine B (2010) Can South Africa be a developmental state? In: Edigheji A (ed) Constructing a democratic developmental state in South Africa: potentials and challenges. HSRC Press, Cape Town, pp 169–182

Fischer-Kowalski M, Swilling M (2011) Decoupling natural resource use and environmental impacts from economic growth. United Nations Environment Programme, Paris

Fleming D (2007) Energy and the common purpose: descending the energy staircase with tradable energy quotas (TEQs). The Lean Economy Connection, London

Fluri TP (2009) The potential of concentrating solar power in South Africa. Energ Pol 37(12):5075–5080

Fofana I, Mabugu R, Chitiga M (2008) Analysing impacts of alternative policy responses to high oil prices using an energy-focused macro–micro model for South Africa. Report prepared for the Financial and Fiscal Commission. FFC, Pretoria

Gagnon N, Hall CAS, Brinker L (2009) A preliminary investigation of energy return on energy investment for global oil and gas production. Energies 2:490–503

GCIS (2007) South Africa yearbook 2006/07. Government Communication and Information System, Pretoria

GCIS (2009) South Africa yearbook 2008/09. Government Communication and Information System, Pretoria

Geels FW (2011) The multi-level perspective on sustainability transitions: responses to seven criticisms. Environ Innov Societal Transitions 1(1):24–40

Gilbert R, Perl A (2008) Transport revolutions: moving people and freight without oil. Earthscan, London

Gilbert R, Perl A (2010) Transportation in the post-carbon world. In: Heinberg R, Lerch D (eds) The post carbon reader: managing the 21st century's sustainability crises. Watershed Media, Healdsburg, CA, pp 1–11

Giller KE, Witter E, Corbeels M, Tittonell P (2009) Conservation agriculture and smallholder farming in Africa: the heretics' view. Field Crop Res 114(1):23–34

Greenpeace (2011) The advanced energy [r]evolution: a sustainable energy outlook for South Africa. Greenpeace International & European Renewable Energy Council, London

Greer JM (2009) The ecotechnic future. New Society Publishers, Gabriola Island, Canada

Guilford MC, Hall CA, Connor PO, Cleveland CJ (2011) A new long term assessment of energy return on investment (EROI) for U.S. oil and gas discovery and production. Sustainability 3(10):1866–1887

Hall CAS (2011) Synthesis to special issue on new studies in EROI (energy return on investment). Sustainability 3:2496–2499

Hall CAS, Benemann JR (2011) Oil from algae? Bioscience 61(10):741–742

Hall CAS, Klitgaard KA (2012) Energy and the wealth of nations: understanding the biophysical economy. Springer, New York

Hall CAS, Ramirez-Pascualli CA (2012) The first half of the age of oil: an exploration of the work of Colin Campbell and Jean Laherrère. Springer, New York

Hall CAS, Balogh S, Murphy DJR (2009) What is the minimum EROI that a sustainable society must have? Energies 2:25–47

Hamilton JD (2009) Causes and consequences of the oil shock of 2007–08. Brookings papers on economic activity. Brookings Institution, Washington, DC

Harkinson J (2009) How the nation's only state-owned bank became the envy of wall street. Mother Jones. http://motherjones.com/mojo/2009/03/how-nation%E2%80%99s-only-state-owned-bank-became-envy-wall-street. Accessed 3 Jun 2011

Hartnady C (2010) South Africa's diminishing coal reserves. S Afr J Sci 106(9/10):1–5

Heckeroth S (2009) Electric tractors. http://www.renewables.com/Permaculture/ElectricTractor. htm. Accessed 18 May 2011

Heinberg R (2004) Powerdown: options and actions for a post-carbon world. New Society Publishers, Gabriola Island, Canada

Heinberg R (2006a) The oil depletion protocol: a plan to avert oil wars, terrorism and economic collapse. New Society Publishers, Gabriola Island, Vancouver

Heinberg R (2006b) Fifty million farmers. New Economics Institute. http://neweconomicsinstitute. org/publications/lectures/heinberg/richard/fifty-million-farmers. Accessed 17 May 2011

Heinberg R (2009) Searching for a miracle: net energy limits and the fate of industrial society, False solution series #4. International Forum on Globalisation, Santa Rosa, CA

Heinberg R (2011) The end of growth: adapting to our new economic reality. New Society Publishers, Gabriola Island, Vancouver

Heinberg R, Bomford M (2009) The food and farming transition: toward a post carbon food system. Post Carbon Institute, Santa Rosa, CA

Hendriks SL (2005) The challenges facing empirical estimation of household food (in)security in South Africa. Dev South Afr 22(1):103–123

Hendriks SL, Msaki MM (2009) The impact of smallholder commercialisation of organic crops on food consumption patterns, dietary diversity and consumption elasticities. Agrekon 48(2): 184–199

Heun MK, de Wit M (2011) Energy return on (energy) invested (EROI), oil prices, and energy transitions. Energ Pol 1185(1):102–118

Hine R, Pretty J, Twarog S (2008) Organic agriculture and food security in Africa. 2007/15. UNEP-UNCTAD Capacity-building Task Force on Trade, Environment and Development, New York

Hirsch RL (2008) Mitigation of maximum world oil production: shortage scenarios. Energ Pol 36(2):881–889

Hirsch RL, Bezdeck R, Wendling R (2005) Peaking of world oil production: impacts, mitigation and risk management. United States Department of Energy, National Energy Technology Laboratory, Washington, DC

Hirsch RL, Bezdek RH, Wendling RM (2010) The impending world energy mess: what it is and what it means to you. Apogee Prime, Burlington, ON

Holm D, Banks D, Schäffler J, Worthington R, Afrane-Okese Y (2008) Potential of renewable energy to contribute to national electricity emergency response and sustainable development. Earthlife Africa, Johannesburg

Hook W (2009) Bus rapid transit: a cost-effective mass transit technology. EM Mag 26–30 June

Höök M, Aleklett K (2010) A review on coal-to-liquid fuels and its coal consumption. Int J Energ Res 34(10):848–864

Hopkins R (2000) The food producing neighbourhood. In: Barton H (ed) Sustainable communities. Earthscan, London, pp 199–215

Hopkins R (2008) The transition handbook. Green Books, Totnes, UK

Howard R (2009) Peak oil and strategic resource wars. Futurist, 18–21 Sept–Oct

Howarth RW, Santoro R, Ingraffea A (2011) Methane and the greenhouse-gas footprint of natural gas from shale formations. Clim Change 106(4):679–690

Hubbert MK (1956) Nuclear energy and the fossil fuels. Proceedings of spring meeting, American petroleum institute drilling and production practice. San Antonio, TX

Hughes JD (2011) Will natural gas fuel America in the 21st century? Post Carbon Institute, Santa Rosa, CA

Hughes JD (2013) Drill, baby, drill: can unconventional fuels usher in a new era of energy abundance? Post Carbon Institute, Santa Rosa, CA

IEA (2005) Saving oil in a hurry. International Energy Agency, Paris

IEA (2012) World energy outlook 2012. International Energy Agency, Paris

IEA (2013) Statistics and balances. International Energy Agency. http://www.iea.org/stats/index.asp. Accessed 30 Jun 2013

IMF (2013) International financial statistics. International Monetary Fund, Washington, DC

INR (2008) Study to develop a value chain strategy for sustainable development and growth of organic agriculture. Investigational report no. IR285. Institute of Natural Resources, Scottsville

Kendall G (2008) Plugged in: the end of the oil age. WWF, Brussels

Kim Y (2013) Nuclear reactions in micro/nano-scale metal particles. Few Body Syst 54(1–4):25–30

Klare MT (2012) The race for what's left: the global scramble for the world's last resources. Macmillan, New York

Korowicz D (2010a) On the cusp of collapse: complexity, energy, and the globalised economy. In: Douthwaite R, Fallon G (eds) Fleeing vesuvius. FEASTA, Dublin, pp 12–39

Korowicz D (2010b) Tipping point: near-term systemic implications of a peak in global oil production. FEASTA, Dublin

Kumhof M, Muir D (2012) Oil and the world economy: some possible futures. IMF working paper WP/12/256. International Monetary Fund, Washington, DC

Lambert J, Hall CAS, Balogh S, Gupta A (2012) EROI of global energy resources: preliminary status and trends. Report prepared by the College of Environmental Science and Forestry, State University of New York. United Kingdom Department for International Development, London

Lane TE (2009) Assessing sustainability and energy efficiency improvement measures in freight transportation. Proceedings of the 28th Southern African transport conference, SATC 6–9 Jul 2009

Leibbrandt MV, Woolard I, Finn A, Argent J (2010) Trends in South African income distribution and poverty since the fall of apartheid. OECD social, employment and migration working papers, No. 101. Organisation for Economic Cooperation and Development, Paris

Leigh J (2008) Beyond peak oil in post globalization civilization clash. Open Geogr J 1:15–24

Litman T (2008) Appropriate response to rising fuel prices. Victoria Transport Policy Institute, Victoria

Lovins A, Datta EK, Bustnes OE, Koomey JG (2005) Winning the oil endgame. Rocky Mountain Institute, Snowmass, CO

Lyne LR, Collins R (2008) South Africa's new cooperatives act: a missed opportunity for small farmers and land reform beneficiaries. Agrekon 47(2):180–197

Mahlalela G (2010) Statement by transport director-general Mr. George Mahlalela at the media briefing on South Africa's rail investment programme. Department of Transport, Pretoria

Mather C (2005) The growth challenges of small and medium enterprises (SMEs) in South Africa's food processing complex. Dev South Afr 22(5):607–622

McKibben B (2007) Deep economy: the wealth of communities and the durable future. Times Books, New York

Modi AT (2003) What do subsistence farmers know about indigenous crops and organic farming? Preliminary experience in KwaZulu Natal. Dev South Afr 20(5):675–684

Mohamed S (2010) The state of the South African economy. In: Daniel J, Naidoo P, Pillay D, Southall R (eds) Growth, resource use and decoupling: towards a "green new deal" for South Africa? Wits University Press, Johannesburg

Mohr SH, Evans GM (2010) Forecasting coal production until 2100. Fuel 88:2059–2067

Murphy D, Hall CAS (2010) Year in review—EROI or energy return on (energy) invested. Ann N Y Acad Sci 1185:102–118

Murphy DJ, Hall CAS (2011) Energy return on investment, peak oil, and the end of economic growth. Ann N Y Acad Sci 1219:52–72

NanoElf Biodiesel (2011) Biodiesel reactors. http://biodiesel.nanoelf.co.za/fullplant.htm. Accessed 1 Mar 2011

National Planning Commission (2012) National development plan 2030: our future—make it work. Presidency of South Africa, Pretoria

National Treasury (2007) Windfall taxes on the liquid fuels industry: response to the task team report on windfall profits in the liquid fuels industry. National Treasury, Pretoria

Niemeyer K, Lombard J (2003) Identifying problems and potential of the conversion to organic farming in South Africa. Unpublished paper. Paper presented at the 41st annual conference of the agricultural economic association of South Africa (AEASA). Pretoria

Njobeni S (2010) Sasol needs state support for Mafutha. Business Day 9 Mar

Nolte M (2007) Commercial biodiesel production in South Africa: a preliminary economic feasibility study. Unpublished thesis. Stellenbosch University, Stellenbosch

North P (2010) Eco-localisation as a progressive response to peak oil and climate change—a sympathetic critique. Geoforum 41:585–594

Odum HT, Odum EC (2011) A prosperous way down. University Press of Colorado, Boulder, CO

Owen NA, Inderwildi OR, King DA (2010) The status of conventional world oil reserves-hype or cause for concern? Energ Pol 38(8):4743–4749

Patzek TW, Croft GD (2010) A global coal production forecast with multi-Hubbert cycle analysis. Energy 35:3109–3122

Pauw KW (2007) Agriculture and poverty: farming for food or farming for money? Agrekon 46(2):195–218

Pegels A (2010) Renewable energy in South Africa: potentials, barriers and options for support. Energ Pol 38(9):4945–4954

PetroSA (2010) Annual report 2010. PetroSA, Cape Town

PetroSA (2012) Annual report 2012. PetroSA, Cape Town

Pfeiffer DA (2006) Eating fossil fuels. New Society Publishers, Gabriola Island, Canada

Philip K (2010) Towards a right to work: the rationale for an employment guarantee in South Africa. TIPS, Pretoria

Pimbert M (2008) Towards food sovereignty: reclaiming autonomous food systems. IIED, London

Pimentel D, Hepperly P, Hanson J, Douds D, Seidel R (2005) Environmental, energetic and economic comparisons of organic and conventional farming systems. Bioscience 55(7):573–582

Prieto PA, Hall CAS (2013) Spain's photovoltaic revolution: the energy return on investment. Springer, New York

Raugei M, Fthenakis V (2012) The energy return on energy investment (EROI) of photovoltaics: methodology and comparisons with fossil fuel life cycles. Energ Pol 45(C):576–582

Rhodes CJ (2009) Oil from algae: salvation from peak oil? Sci Prog 92(1):39–90

Roelf W (2012) S. Africa sees $258 mln ethanol plant by 2014. http://www.reuters.com/article/2012/02/20/ozabs-safrica-biofuels-idAFJOE81J05I20120220. Accessed 5 Feb 2013. Reuters 20 Feb

Rosset P (1999) Small is bountiful. Ecologist 29(8):452–456

Rubin J (2009) Why your world is about to get a whole lot smaller: oil and the end of globalisation. Random House, New York

Rustomjee Z, Crompton R, Maule A, Mehlomakulu B, Steyn G (2007) Possible reforms to the fiscal regime applicable to windfall profits in South Africa's liquid fuel energy sector, with particular reference to the synthetic fuel industry. Windfall Tax Task Team, Pretoria

Rutledge D (2011) Estimating long-term world coal production from S-curve fits. Int J Coal Geol 85:23–33

SANRAL (2010) Annual report 2010. South African National Road Authority Limited, Johannesburg

SAPIA (2013) Annual report 2012. South African Petroleum Industry Association, Johannesburg

SARB (2013) Quarterly bulletin. South African Reserve Bank, Pretoria

Sasol (2010) Sasol facts 2010. Sasol Limited, Johannesburg

SAWEA (2010) Our sustainable future: the case for wind energy, proposed integrated resource plan submission by the wind energy industry. South African Wind Energy Association, Johannesburg

Schiller PL, Bruun EC, Kenworthy JR (2010) An introduction to sustainable transportation: policy, planning and implementation. Earthscan, London

Schulschenk J (2010) Benefits and limitations of local food economies to promote sustainability: a Stellenbosch case study. Unpublished thesis. Stellenbosch University, Stellenbosch

Shafirovich E, Varma A (2009) Underground coal gasification: a brief review of current status. Ind Eng Chem Res 48(17):7865–7875

Cap and Share (2011) Cap and share. http://www.capandshare.org/index.html. Accessed 13 May 2011

Sharp J (2008) "Fortress SA": xenophobic violence in South Africa. Anthropol Today 24(4):1–3

Shuman M (1997) Going local: creating self reliant communities in a global age. Free Press, New York

Simms A (2008) Nine meals from anarchy: oil depletion, climate change and the transition to resilience, Schumacher lecture, 2008. New Economics Foundation, London

Sorrell S, Miller R, Bentley R, Speirs J (2010a) Oil futures: a comparison of global supply forecasts. Energ Pol 38(9):4990–5003

Sorrell S, Speirs J, Bentley R, Brandt A, Miller R (2010b) Global oil depletion: a review of the evidence. Energ Pol 38(9):5290–5295

Spalding-Fecher R, Matibe D (2003) Electricity and externalities in South Africa. Energ Pol 31(8):721–734

Srivastava Y, Widom A, Larsen L (2010) A primer for electroweak induced low-energy nuclear reactions. Pramana J Phys 75(4):617–637

StatsSA (2012) Income and expenditure survey 2010/2011. Statistics South Africa, Pretoria

StatsSA (2013) Statistics online. Statistics South Africa. http://www.statssa.gov.za/. Accessed 20 Aug 2013

Strahan D (2007) The last oil shock. John Murray, London

Tainter JA (1988) The collapse of complex societies. Cambridge University Press, Cambridge

Taleb NN (2007) The black swan: the impact of the highly improbable. Random House, New York

The Presidency (2008) Towards a fifteen year review. The Presidency, Republic of South Africa, Pretoria

The Presidency (2009) Development indicators 2009. The Presidency, Republic of South Africa, Pretoria

The Presidency (2010) Strategic plan 2010/11 to 2012/13. The Presidency, Republic of South Africa, Pretoria

Thornton A (2008) Beyond the metropolis: small town case studies of urban and peri-urban agriculture in South Africa. Urban Forum 19:243–262

TIPS (2010) Community work programme annual report 2009/10. Department of Cooperative Governance, Pretoria

Trollip H, Marquard A (2010) Prospects for renewable energy in South Africa. Heinrich Boll Stiftung, Cape Town

Trollip H, Tyler E (2011) Is South Africa's economic policy aligned with our national mitigation policy direction and low carbon future: an examination of the carbon tax, industrial policy, new growth path and integrated resource plan. National Planning Commission, Pretoria

Turok I, Hunter R, Robinson B, Swilling M, van Ryneveld P (2011) Towards resilient cities. South African Cities Network, Johannesburg

UNDP (2010) Human development report 2010. United Nations Development Programme, New York

van den Bergh JCJM, Truffer B, Kallis G (2011) Environmental innovation and societal transitions: introduction and overview. Environ Innov Societal Transitions 1(1):1–23

Vanderschuren M, Jobanputra R, Lane T (2008) Diminishing global oil supply: potential measures to redress the transport impacts. J Energ South Af 19(3):20–29

Vink N, van Rooyen J (2009) The economic performance of agriculture in South Africa since 1994: implications for food security. Development planning division working paper series no. 17. Development Bank of Southern Africa, Midrand

Wakeford JJ (2012) Socioeconomic implications of global oil depletion for South Africa: vulnerabilities, impacts and transition to sustainability. PhD dissertation. Stellenbosch University, Stellenbosch, South Africa

Westley F, Olssen P, Folke C, Homer-Dixon T, Vredenburg H, Loorbach D, Thompson J, Nilsson M, Lambin E, Sendzimir J, Banarjee B, van der Leeuw S (2011) Tipping towards sustainability: emerging pathways of transformation. Third Nobel laureates symposium on global sustainability: transforming the world in an era of global change. Stockholm. Accessed 16–19 May

Winkler H, Jooste M, Marquard A (2010) Structuring approaches to pricing carbon in energy- and trade-intensive sectors: option for South Africa. Putting a price on carbon: economic instruments to mitigate climate change in South Africa and other developing countries. University of Cape Town, Cape Town

Woodson M, Jablonowski CJ (2008) An economic assessment of traditional and cellulosic ethanol technologies. Energ Sourc B Energ Econ Plann 3(4):372–383

World Bank (2013) World development indicators. World Bank. http://databank.worldbank.org/ddp/home.do. Accessed 4 Mar 2013

Zittel W, Zerhusen J, Zerta M, Arnold N (2013) Fossil and nuclear fuels—the supply outlook. Energy Watch Group, Berlin

Index

A

Agriculture
  commercial, 49–51, 55, 60
  input costs, 51–54, 57, 58, 60
  low-till, 56
  organic, 50, 56–60, 91, 92
  sustainable, 55, 57, 58, 92
  urban, 60, 61, 98, 108
Arable land, 26, 27, 33, 49, 56, 99

B

Bicycles, 44, 47, 74, 76, 102
Biofuels, 11, 22, 26–27, 30, 33, 74, 98
Biogas, 22, 27, 33, 60, 103
Biomass, 8, 11, 27, 28, 64, 74, 98, 103
Bus rapid transit (BRT), 36, 38, 40, 46,
    47, 102

C

Cap and Dividend (C&D), 89
Cap and Share (C&S), 86, 88–90
Car-pooling, 41–42, 102, 104
Coal, 4, 8, 11, 13, 16, 18, 20, 22–26, 28,
    30, 31, 36, 37, 50, 51, 64–66,
    68, 69, 72, 74, 77, 84, 89,
    95–97, 105, 106
Coal-to-liquids (CTL), 13–15, 21–23, 30–32,
    64, 65, 72, 95–97, 100
Cooperatives, 32, 58, 90, 91, 93, 98, 99
Crime, 42, 44, 82–84
Critical systems, 84, 85, 94
Current account deficit, 67

D

Debt
  household, 67
  public, 68, 70
Depletion, 1, 2, 6, 20, 30, 40, 67, 70, 71, 76,
    87, 89, 94, 99, 106

E

Eco-driving, 41, 103
Economic barriers, 106
Electricity, 8, 11, 18, 20, 22, 23, 25, 27–29,
    31–33, 36, 37, 43, 45–48, 50, 51,
    64, 69, 71, 73, 75, 81, 84, 88, 89,
    97, 98, 102–105, 108
Energy
  consumption, 8, 11, 12, 18–20, 36, 37,
    50, 88
  efficiency, 21, 31, 32, 41, 43, 44, 46, 47,
    64, 71, 72, 74, 77, 79, 99, 106, 107
  intensity, 8, 51, 63–65
  nuclear, 4, 33
  renewable, 20, 28, 29, 61, 71, 73–76, 89,
    91, 96, 97, 102, 103, 105, 106, 110
  solar, 74
  wind, 29
Energy return on investment (EROI), 3, 5, 23,
    26, 29, 32, 74, 109

F

Farming
  agroecological, 55–61, 108
  subsistence, 49–51